Palgrave Studies in the Theory and History of Psychology

Palgrave Studies in the Theory and History of Psychology publishes scholarly books that use historical and theoretical methods to critically examine the historical development and contemporary status of psychological concepts, methods, research, theories, and interventions. The books in the series are characterised by an emphasis on the concrete particulars of psychologists' scientific and professional practices, together with a critical examination of the assumptions that attend their use. These examinations are anchored in clear, accessible descriptions of what psychologists do and believe about their activities. All the books in the series share the general goal of advancing the scientific and professional practices of psychology and psychologists, even as they offer probing and detailed questioning and critical reconstructions of these practices.



More information about this series at
http://www.palgrave.com/gp/series/14576

Kieran C. O'Doherty · Lisa M. Osbeck ·
Ernst Schraube · Jeffery Yen
Editors

# Psychological Studies of Science and Technology

*Editors*
Kieran C. O'Doherty
Department of Psychology
University of Guelph
Guelph, ON, Canada

Lisa M. Osbeck
Department of Psychology
University of West Georgia
Carrollton, GA, USA

Ernst Schraube
Department of People and Technology
Roskilde University
Roskilde, Denmark

Jeffery Yen
Department of Psychology
University of Guelph
Guelph, ON, Canada

Palgrave Studies in the Theory and History of Psychology
ISBN 978-3-030-25307-3 ISBN 978-3-030-25308-0 (eBook)
https://doi.org/10.1007/978-3-030-25308-0



This Palgrave Macmillan imprint is published by the registered company Springer Nature Switzerland AG
The registered company address is: Gewerbestrasse 11, 6330 Cham, Switzerland

# Contents

# Notes on Contributors

**Natasha Bharj** is a graduate student at the University of Kansas. She locates her scholarship at the intersection of decolonial feminism and critical cultural psychology. She is primarily interested in socio-cultural constructions of sexuality and sexual violence, particularly in the contexts of coloniality and narratives of development. She also conducts research on representations of history and national identity, and is invested in the scholarship of teaching & learning.

**Dr. Markus Brunner** is a Lecturer at the Sigmund Freud Private University in Vienna, board member of the Gesellschaft für psychoanalytische Sozialpsychologie ["Society for psychoanalytic social psychology"], and co-editor of the journals "Freie Assoziation. Zeitschrift für psychoanalytische Sozialpsychologie" [Free Associations] and "Psychologie und Gesellschaftskritik" [Psychology and Social Critique]. He has written and published broadly on psychoanalysis and critical theory with a special focus on trauma theory, anti-semitism, and the emotional legacies of Nazi Germany.

**Niklas Alexander Chimirri** is an Associate Professor in Social Psychology of Everyday Life at the Department of People and Technology at Roskilde

University, Denmark. Both his teaching and research explore the relevance of communication technology for children's and adults' conduct of everyday life, with a focus on technology's ethical implications for designing collaborative research across generational thresholds. The aim is to develop a praxis-philosophical, socio-material psychology of everyday life that interrelates audience research, participatory design studies and intergenerational childhood research. He currently investigates young children's and adults' relational care practices in digitalized times.

**Joshua W. Clegg** is an Associate Professor of psychology at John Jay College of Criminal Justice, CUNY and a member of the faculty in the Critical Social and Personality Psychology doctoral program at The Graduate Center, CUNY. He has published widely on history, methods, and epistemology in the social sciences.

**Shannon Cunningham** has a Ph.D. in Applied Social Psychology, with a specialization in evaluation and qualitative methods. She works as a Senior Research Evaluator and Senior Operations Consultant at Alberta Innovates, a provincially funded corporation that supports research and innovation development in the core sectors of health, environment, energy, food and fibre, and ecosystem biodiversity. Shannon supports the corporation through corporate and strategic planning, performance management and measurement activities, managing and conducting evaluations, impact assessments and narratives, and continuous improvement activities. She also freelances as a qualitative methods expert on quality improvement health communications research projects.

**Ass-Prof. Dr. Katharina Hametner** is a psychologist who is working at Sigmund Freud University Vienna. She wrote her Ph.D. on experiences of everyday racist practices in the field of anti-Muslim discourses at the University of Vienna. Her research and teaching expertise lies within the fields of social psychology, with special interests in racism and migration, qualitative methods and philosophy of science. She is co-editor of the journal "Psychologie und Gesellschaftskritik" (psychology and social criticism).

**Peter Hegarty** is a Professor of Psychology at the University of Surrey where he directs the Social Emotions and Equality in Relations research group SEER (www.surrey.ac.uk/seer), and delivers the curriculum on Conceptual and Historical Issues in Psychology (CHIP). His research interests focus on social psychology, psychology's history, and gender and sexuality, and are currently focused on public understanding of the medicalization and demedicalization of intersex. He is the author of *Gentlemen's Disagreement: Alfred Kinsey, Lewis Terman and the Sexual Politics of Smart Men* (Chicago, 2013) and *A Recent History of Lesbian and Gay Psychology: From Homophobia to LGBT* (Routledge, 2017).

**Katherine Hubbard** is a feminist academic at the University of Surrey. She specializes in the history of Psychology; gender and sexuality studies; and socio-historical methods. Her work centralizes on disrupting normative judgements about the past and demonstrates the how useful queer feminist histories are to the social sciences and beyond. Her book *Queer Ink: A Blotted History Towards Liberation* does this in relation to the Rorschach ink blot test and its relationship with gay liberation movements inside and outside of Psychology.

**Amanda Jenkins** received her Ph.D. in Applied Social Psychology from the University of Guelph in Ontario, Canada. Her research interests are health, feminist theory, and the body. Her publications include Clean and fresh: Understanding women's use of vaginal hygiene products (*Sex Roles*, 2018). She is currently a Research Coordinator for a cross-national AGE-WELL funded project titled Older Adults' Active Involvement in Ageing & Technology Research and Development (OA-INVOLVE).

**Ines Langemeyer** is a Professor at the Karlsruhe Institute of Technology for the research on learning and instruction, the philosophy of education and vocational education. She obtained her Diplom in psychology at the Free University of Berlin and her Ph.D. (Dr. Phil.) at the Hamburg Helmut-Schmidt-University. Main research interests are research- or inquiry-based forms of teaching and learning, the scientification of work, historical epistemological approaches to science, the development of cooperative competence, and the conditions of high-performance workplaces on the labor market.

**Nathalie Lovasz** is founding partner and practicing as a clinical psychologist at the Toronto DBT Centre in Toronto, Canada. She holds a Ph.D. from Simon Fraser University and has published on various topics at the intersection of history and theory of psychology and clinical practice.

**Nancy J. Nersessian** is Regents' Professor (Emerita), Georgia Institute of Technology. She currently is Research Associate, Department of Psychology, Harvard University. Her research focuses on the creative research practices of scientists and engineers, especially how modeling practices lead to fundamentally new ways of understanding the world. This research has been funded by NSF and NEH. She is a Fellow of AAAS, the Cognitive Science Society, and a Foreign Member of the Royal Netherlands Academy of Arts and Sciences. Her numerous publications include *Creating Scientific Concepts* (MIT, 2008, Patrick Suppes Prize in Philosophy of Science, 2011) and *Science as Psychology: Sense-Making and Identity in Science Practice* (with L. Osbeck, K. Malone, W. Newstetter, Cambridge, 2011, APA William James Book Prize, 2012).

**Kieran C. O'Doherty** is an Associate Professor in the department of psychology at the University of Guelph, where he directs the Discourse, Science, Publics research Group. His research focuses on the social and ethical implications of science and technology. In this context, he has published on such topics as vaccines, human tissue biobanks, the human microbiome, salmon genomics, and genetic testing. Kieran's research also emphasizes public engagement on science and technology. In this regard, he has designed and implemented public deliberations in which members of the public engage in in-depth discussion about ethical aspects of science and technology and collectively develop recommendations for policy. Kieran's research has been funded by the Canadian Institutes of Health Research, The Office of the Privacy Commissioner of Canada, the Ontario Ministry of Research & Innovation, Genome Canada and Genome British Columbia. He is editor of *Theory & Psychology*.

**Lisa M. Osbeck** is a Professor of Psychology at the University of West Georgia. Her interests include historical and contemporary theories of cognition, philosophy of science, methodology, and problems of interdisciplinary research. Books include *Science as Psychology: Sense-Making* *and Identity in Science Practice* (with N. Nersessian, K. Malone, & W. Newstetter, Cambridge, 2011), awarded the William James Book Award from the American Psychological Association in 2012, and *Values in Psychological Science: Re-Envisioning Epistemic Priorities* at a New Frontier (Cambridge, 2019). She is also co-editor of *Rational Intuition: Philosophical Roots, Scientific Investigations* (Cambridge, 2014), with Barbara Held. Lisa is a Fellow of the American Psychological Association and is President-Elect of the Society for General Psychology (APA Division 1). She is an associate editor for *Qualitative Psychology* and is on the editorial boards of six additional journals, including *American Psychologist*.

**Michael Pettit** is an Associate Professor at York University. His primary affiliation with the Historical, Theoretical, and Critical Studies area within the Department of Psychology. He works at the intersection of critical psychology, Science & Technologies Studies, and the history of the human sciences.

**Nora Ruck** is an Assistant Professor of Psychology at the Sigmund Freud Private University in Vienna, where she co-coordinates the master program social psychology and psychosocial practice. Her research interests concern the relations between psychology and social inequalities as well as social movements, the history of psychology, feminist epistemology, and critical psychology. She currently directs the research project "The psychological is political. A recent history of feminist psychology in Vienna, 1972–2000" (FWF P 31123-G29), which studies the history of feminist psychologies in Austria. She is recipient of the Austrian state prize for excellent university teaching (ars docendi) for a transnational (Austria/U.S.) seminar on "Holocaust consciousness" (with Bernadette Wegenstein & Markus Brunner) as well as co-editor of "Psychologie & Gesellschaftskritik" and associate editor of *Awry—Journal of Critical Psychology*.

**Alexandra Rutherford** is a Professor of psychology at York University in Toronto where she is a member of the Historical, Theoretical, and Critical Studies of Psychology graduate program. She studies the relationships between feminist psychology and society to understand how feminist science and practice can contribute to positive social change. She has examined the role of sexual assault surveys in mobilizing anti-violence activism and policy, and the influence of neoliberalism and postfeminism in shaping gendered discourses of agency and empowerment. She received the Association for Women in Psychology Distinguished Publication Award in 2012 for her co-edited volume *Handbook of International Feminisms: Perspectives on Psychology, Women, Culture, and Rights*, and the Distinguished Member Award from the Section on Women and Psychology of the Canadian Psychological Association in 2011. She is the founder and director of the Psychology's Feminist Voices oral history and digital archive project.

**Ernst Schraube** is a Professor of Social Psychology of Technology in the Department of People and Technology at Roskilde University, Denmark. His research focuses on the social and political implications of modern technology in everyday life and he is currently working on a project on the significance of digital technologies in students' learning and conduct of everyday life. He studied at the Free University Berlin from where he received his diploma in psychology and his Ph.D. Among his books are *Auf den Spuren der Dinge: Psychologie in einer Welt der Technik* (Argument) and the co-edited volumes *Psychology from the Standpoint of the Subject: Writings of Klaus Holzkamp* (Palgrave Macmillan), *Psychology and the Conduct of Everyday Life* (Routledge). The 1998–1999 academic year he spent as a visiting research scholar in the Department of Science and Technology Studies at Rensselaer Polytechnic Institute in Troy, New York, USA. More information under: www.ruc.dk/~schraube.

**Kathleen Slaney** is a Professor of Psychology at Simon Fraser University. Her research interests and expertise span several areas, including historical and conceptual analysis of the methods and methodologies of psychological science, philosophy of psychology and related disciplines, and theoretical and applied psychometrics. She is a Fellow of the American Psychological Association, author of *Validating Psychological Constructs* (Palgrave Macmillan, 2017), and co-editor of *A Wittgensteinian Perspective on the Use of Conceptual Analysis in Psychology* (Palgrave Macmillan, 2013) and *The Wiley Handbook of Theoretical and Philosophical Psychology* (Wiley Blackwell, 2015).

**Estrid Sørensen** with a background in Psychology from the University of Copenhagen, has focussed her academic life on studying how things, technologies and other objects engage in everyday lives. She is the author of *The Materiality of Learning: Technology and Knowledge in Educational Practice* (Cambridge, 2009) which is based on micro-analytic studies in a primary school context in Denmark. She has studied how objects shape everyday lives in a variety of other contexts such as educational assessment exercises, computer game activities, and psychological knowledge practices. She is a former Alexander von Humboldt awardee and holds currently a Professorship in Cultural Psychology and Anthropology of Knowledge at the department of Social Science at Ruhr-University in Bochum. She has also taught at departments of anthropology, education, and sociology in Denmark and Germany.

**Henderikus J. Stam** is a Professor of Psychology at the University of Calgary, where he teaches courses in the history and theory of psychology as well as clinical psychology. His recent scholarly interests have focused on contemporary theoretical problems in psychology and the historical foundations of twentieth-century psychology on which he has published numerous articles and book chapters. Founding editor of *Theory & Psychology* (now Editor Emeritus), he is a former president of APA's Division 26 (Society for the History of Psychology) and of Division 24 (Society for Theoretical and Philosophical Psychology).

**Donna Tafreshi** is an Adjunct Professor in the Psychology Department at Simon Fraser University. Her background and scholarship are in the areas of research methodology, developmental psychology, and the history and philosophy of science.

**Thomas Teo** is a Professor of psychology in the Historical, Theoretical, and Critical Studies of Psychology Program at York University (Toronto, Canada). He has been active in the advancement of theoretical, critical, and historical psychology throughout his professional career. His research has been meta-psychological to provide a more reflexive understanding of the foundations, trajectories, and possibilities of human subjectivity. He considers his research program contributing to the psychological humanities.

**Sapphira R. Thorne** is a Postdoctoral Research Associate at Cardiff University, UK. Her research examines the role of heteronormativity in the construction of love and romantic relationships. She is also more broadly interested in relationships, gender and sexuality, intergroup discrimination, social norms, and mindfulness. Recent publications include "Equality in theory: From a heteronormative to an inclusive psychology of romantic love" (*Theory & Psychology*, 2019).

**Charlie A. Wu** is a graduate student currently enrolled in the Historical, Quantitative, and Theoretical Psychology department at Simon Fraser University. His interests include the history of moral psychology, qualitative and quantitative methodology in the social sciences, hermeneutic investigation in psychology, and critical psychology.

**Jeffery Yen, Ph.D.** is an Associate Professor in Applied Social Psychology at the University of Guelph. His teaching and research interests centre on critical history and theory of psychology, with a focus on locating psychological technoscience in its social, cultural, and historical context. He is currently researching the history of clinical psychology in South Africa, as well as continuing research on the performativity of methods used in psychological science. He has co-edited the book *Theoretical Psychology: Global Transformations and Challenges*, is currently Book Review Editor for the journal *Theory & Psychology*, and serves on the editorial board for the journal *Awry: Journal of Critical Psychology*.

# List of Figures

# 1

# Introduction: Psychological Studies of Science and Technology

**Kieran C. O'Doherty, Lisa M. Osbeck, Ernst Schraube and Jeffery Yen**

Recognition of the centrality of science and technology to almost all domains of human activity has given rise to subdisciplines such as philosophy of science, philosophy of technology, sociology of knowledge, and history of science and technology. Recent years are witness to the emergence of prominent interdisciplines such as Science and Technology Studies (STS) that offer systematic investigation of the

K. C. O'Doherty (✉) · J. Yen
University of Guelph, Guelph, ON, Canada
e-mail: odohertk@uoguelph.ca

J. Yen
e-mail: jyen@uoguelph.ca

L. M. Osbeck
Department of Psychology, University of West Georgia, Carrollton, GA, USA
e-mail: losbeck@westga.edu

E. Schraube
Department of People and Technology, Roskilde University, Roskilde, Denmark
e-mail: schraube@ruc.dk



K. C. O'Doherty et al. (eds.), *Psychological Studies of Science and Technology*, Palgrave Studies in the Theory and History of Psychology,
https://doi.org/10.1007/978-3-030-25308-0_1

dynamic relationships between science, technology, and human life. Somewhat surprisingly, the discipline of psychology has been comparatively marginal to STS and related research. There are three main markers of its marginality. First, with some notable exceptions, psychology has not been prominent as an *object of study* within STS, which instead has focused more on an investigation of the natural sciences. Of course, the critical scrutiny of psychological science has contributed to the emergence of subdisciplines within psychology, including theoretical psychology, critical psychology, and feminist psychology, but interface between these efforts and STS have been minimal. Second, psychology as a scholarly discipline with a distinct analytic lens has been underrepresented in the interdisciplinary and multidisciplinary efforts of STS. Third, although we may identify a literature associated with the label *psychology of science*, that is, the psychological study of science and technology, as a focus of research it has arguably remained quite marginal to the discipline of psychology itself.

Previous attempts to articulate and develop a psychology of science and technology have not had much influence in the larger STS community. As we argue in more detail below, one reason for this is that these previous attempts have tended to have a rigid commitment to an unreflexive epistemology that does not allow for historical, critical, constructionist, qualitative, and theoretical scholarship. In contrast, STS quite explicitly draws on a plurality of disciplines and scholarly approaches, and very notably interrogates the limits of the scientific onto-epistemologies of western modernity (e.g., Latour, 1993b). Certainly, important historical work on cognitive practices of scientists using interpretive methods helped to carve out a distinct subspecialty of cognitive studies of science, beginning in the 1980s (e.g., Nersessian, 1984; Tweney, 1989), and there has been a great deal of important critical historical analysis of scientific practices specifically relevant to psychological science (e.g., Danziger, 1994; Gergen, 1973; Koch & Leary, 1992; Morawski, 1988). More recently, qualitative methods have been used to explore themes broadly relevant to social and cognitive processes in science on the part of a growing empirical philosophy of science community (see Wagenknecht, Nersessian, & Andersen, 2015). However, the relation of these efforts and the interpretive methods used to a broader project of psychology of science have not been articulated in the formal calls for a subfield to be explicitly associated with the label "psychology of science" (e.g., Feist & Gorman, 2013). Presumably, then, psychological contributions to STS, or studies of the psychological dimensions of science, if they are to be compatible with the spirit and practice of STS, would need to exhibit a reflexivity and plurality of scholarly frameworks, approaches, and methods. This book is aimed precisely at addressing this need.

By way of introduction, we situate our project in the broader context of (1) psychological studies of science and (2) psychological studies of technology, and (3) psychological contributions to STS.

## Psychological Studies of Science

Any story told about the origin and development of "psychology of science" reflects assumptions about the nature of history, psychology, and science, and about their interrelations, thus complicating the task of situating our current project. An important first task is to distinguish (1) a formal scholarly specialty identified by the name "psychology of science" and (2) empirical and theoretical analysis of science not explicitly identified as "psychology of science" that nevertheless addresses broadly psychological concerns. In the case of the latter, there are several tributaries, each of which in their own right would require extensive review to even begin to do them justice. We can consider these the narrow and broad senses of psychology of science, respectively. Greg Feist makes this distinction in his effort to summarize and schematize the enormous set of contributions that could be considered part of the overall project of psychology of science. His work, singly (2006) and in combination with Gorman (e.g., Feist & Gorman, 1998, 2013), offers comprehensive and detailed historical survey, as does Simonton (1988). Campbell (1989) and Houts (1989) conducted comprehensive historical analyses of the philosophical underpinnings of psychology of science in the pioneering *Psychology of Science: Contributions to Metascience* (Gholson, Shadish, Neimeyer, & Houts, 1989). Here we are able only to offer a sketch of both the broad and narrow sense, drawing in part from Feist's history but also adding some thoughts of our own. Our sketch is intended only to invite further reading and conversation, especially around the question of the appropriate demarcation and units of analysis of which a psychology of science is inclusive. We should note regretfully that our summary covers only North American and European psychology and science, owing only to our limitations in space and expertise.

## Philosophical Studies of Science

Most fundamentally, it is important to first recognize that epistemology itself as historically unfolded is a kind of psychology of science. From Plato forward disagreement concerning the origin of knowledge focused on which human capacity can be trusted as the basis for knowledge claims, with the primacy of intellectual powers pitted against the rock bottom importance of sensory channels. The emergence of modern science is also defended philosophically on grounds that are at least in some aspects "psychological", if by psychological we would include the forms of bias to which our everyday reasoning is prone. For example, against the rapid expansion of scientific discovery in the late sixteenth and early seventeenth centuries, Francis Bacon based the call for a "new instrument" for knowledge acquisition on what are essentially psychological grounds: four categories of "idols" of mind that serve to skew unaided observation and sense-making and thereby lead to hasty or erroneous conclusions. The idols as described represent psychological categories or kinds that would not be out of step in a contemporary curriculum: cognitive mechanisms ("the tribe"), disposition and habit ("the cave"), ambiguities stemming from language and communication ("the marketplace"), and deeply engrained culturally normative ideation ("the theater"). Bacon's call for a more systematic method of inquiry proceeds not from analysis of the nature of the objects investigated so much as on the features (the limitations) of the inquiring subject (Bacon, *Novum Organum*, 1937/1620). One might note also that historians of psychology have analyzed the psychologically thematic content of many other early modern philosophers who were concerned with the grounding of scientific knowledge (e.g.,

Descartes, Hobbs, Locke, Hume, Kant); (see Leahey, 2017; Pickren & Rutherford, 2010; Robinson, 1995).

These developments in philosophy, aimed at understanding the nature and limits of scientific discovery, are not themselves designed to produce new knowledge of the world, new scientific insights, that is. Rather, they concern the human abilities that enable or constrain the construction and advancement of knowledge. As James put it, "Locke, Hume, Berkeley, Kant, Hegel, have all been utterly sterile, so far as shedding any light on the details of nature goes… The satisfactions they yield to their disciples are intellectual, not practical…" (1987/1907, p. 568). Overlooking for the present purposes the many differences between the thinkers James links together to make his argument, their views may be considered psychological treatises on the nature of scientific reasoning—reflections on science as an object—thus psychological studies of science—in a broad but important sense.

If we were to conform with the thinking that psychology is distinguished by empirical methods, that is, by the systematic collection and analysis of data rather than philosophical reflection, we could trace the origins of a rudimentary psychology of science to the experimental study of the "personal equation," astronomer's awareness and experimental investigation of individual differences in estimates of the movement of planetary bodies among astronomers using the same telescope, dubbed by Schaffer as "the personality problem" (see Schaffer, 1988). Yet the inclusion of data alone cannot be specific to mark the pursuit as psychological. It remains difficult to distinguish a psychology of science from the many debates about the nature of science taking place during the late eighteenth and nineteenth centuries, including heated exchange concerning the nature and possibility of a psychological science, much of which drew from systematic collection of data relating to perceptual processes (e.g., Fechner, 1860; von Helmholtz, 1995/1868). Discussion of the mind's contribution to the context of perception, the potential for the senses to deceive us, the relation of scientific inquiry to goals and values all straddle the philosophical and psychological realms and converge around the contribution of the observer to what is observed (e.g., Mach, 1896).

Similarly, nineteenth and early twentieth centuries scholarship on the nature of science, including the articulation of "pragmatism" by James himself (1987/1907), and encompassing the insights of Whewell (1847), Brentano (1874), Mill (1843), Peirce (1878), Dewey (1938), Husserl (1954/1936), and Wittgenstein (1953), to name only a few examples, provides ample evidence that psychological themes of one kind or another (including, importantly, our use of language) infiltrate the philosophical study of science as a form of human activity and achievement. More recent philosophy of science varies in the extent to which psychological subject matter receives emphasis (see Douglas, 2009). Polanyi's *Personal Knowledge* is explicit in foregrounding the importance of passion, commitment, resilience, and unconscious processing to the very possibility of scientific advance (1974/1958), in calling discovery "an extremely delicate and personal art" (1964/1946, p. 34). But even within the analytic tradition, most philosophers of science demonstrated willingness to admit that at the level of "discovery" at least (i.e., distinct from the verification or "test" of an idea), scientific reasoning requires more than following a recipe, thus inviting discussion of the extent to which disposition, cognitive style, development, and sociality might have a bearing on the conditions that make science possible (Reichenbach, 1938). To the extent that the questions posed about discovery concern the nature of human creativity and insight, the overlap with psychological studies of creative and productive thought processes is obvious. Thus Popper famously levied a "sharp distinction" between the contexts of discovery and justification, and considered the former "of great interest to empirical psychology" but "irrelevant to logical analysis" (2002/1959, p. 8).

## Cognitive Studies of Science

The idea that the study of scientific discovery or insight is a distinctly psychological project and not a philosophical one seems to have inspired much creative energy on the part of psychology, some of it reflecting methodological innovation. As an example, Max Wertheimer conducted an extensive and intricate interview with Albert Einstein targeted at understanding the process of thinking to which Einstein credited his most influential insights, and those that enabled his outline of the general theory of relativity (Wertheimer, 1981/1948).

Renewed interest in cognition among postwar empirical psychologists launched an interdisciplinary subfield of cognitive studies of science later in the twentieth century. That is, the growing interest in cognitive structures and information processing invited a slew of new experimental and theoretical investigations of reasoning that had relevance for understanding scientific reasoning in particular. Herbert Simon (1981/1966), for example, argued from data related to symbol processing and selective forgetting mechanisms to an explanation of the processes of incubation and illumination common to both everyday problem-solving and leaps of creative scientific insight. Philip Johnson-Laird (1983), drawing on the earlier insights of Kenneth Craik (1943) characterized human thinking in terms of the construction and manipulation of mental models of various types and levels of complexity. This work in turn inspired a generation of cognitive scientists as well as philosophers of science to dedicate their efforts to better understand the cognitive basis of successful scientific reasoning, including the construction and use of various kinds of models (Gentner & Stephens, 1983; Giere, 1992; Gooding, 1985; Gorman & Carlson, 1990; Nersessian, 1984, 1992; Tweney, 1989; Tweney, Doherty, & Mynatt, 1981). This work was inherently interdisciplinary: their concepts and methods drew from different disciplines—cognitive psychology, of course, but also historical document analysis and philosophical methods, in the process of which these scholars contributed not only to deeper understanding of scientific reasoning but also broadened the scope of cognitive science to include fine-grained analysis of historical case studies.

## Sociological Studies of Science

At the same time, seemingly in parallel, sociological studies of science offered competing accounts of both discovery and verification processes. In the 1960s, new metatheoretical accounts of scientific progress galvanized philosophical discourse around the nature of discovery *and* verification (justification), offering what amount to largely sociological rather than logical accounts of scientific progress (e.g., Feyerabend, 1975; Foucault, 1966; Kuhn, 1962; Rorty, 1979). Instead of cognitive architecture, sociocultural arguments, including "social epistemology," depicted science as a set of conventions and organizational structures—deeply entrenched habits of thought and linguistic practice, socially shared representations, and negotiation of conceptual and methodological norms—rules within communities, including rules for testing hypotheses (Bloor, 1975; Fuller, 2002; Latour, 1987; Latour & Woolgar, 1986). That the communities in question have agendas that are at base politically and economically driven, concerned with power and subversion, with interests that are instantiated in the institutions that support scientific knowledge production is an important aspect of accounts offered.

Similarly, feminist studies of science illuminated the role of power imbalances and intersubjective relations in the construction of scientific knowledge, including deep biases infiltrating the criteria for judging the worth of ideas, contributing to entrenched discriminatory practices that limit and devalue the contribution of women to knowledge production (Eagly, 1987; Fox-Keller, 1985; Haraway, 1985; Harding, 1986). Cultural studies of science and critical race theory, drawing inspiration from Frantz Fanon (1967/1952), the broader tradition of critical theory (e.g., Adorno & Horkheimer, 2002/1944; Foucault, 1975; Teo, 2015), and from historical documentation of racialized practices in science have offered much insight into the means by which myriad forms of prejudice affect who is entitled to make and use knowledge, who is excluded, and who is harmed by its production (Guthrie, 2004; Malone & Barbarino, 2009; Sinclair, 2004). These influences have also had a profound impact on theoretical psychologists' critical analyses of the politicized nature of their discipline's own knowledge producing practices (e.g., Burman, 2016; Gergen, 2000; Teo, 2017), all of which must be recognized as part of broader interdisciplinary subject matter of a psychology of science.

## Psychological Studies of Science

Despite the important understandings of science arising from both cognitive science and sociocultural analysis, there remain some distinctly or at least importantly psychological contributions to the study of science. Feist and Gorman locate the "very first inklings" of a psychology of science in the late nineteenth century scuffling between historian Alphonse de Candolle, who emphasized social facts such as religious affiliation that showed an association with scientific achievement (Feist & Gorman, 2013, p. 4). Francis Galton's analysis of survey data from 180 scientists to make a wobbly argument for the hereditary basis of scientific capacities (see Fancher, 1983). Somewhat later, Bachelard (2002/1938) illuminated links between poetry and epistemology through analysis of analogy, metaphor, and the role of imagination in scientific thinking. Similarly of note is D. L. Watson's *Scientists Are Human*, which asserts "the scientist" to be central to an understanding of science: "Central to any estimate of the nature of scientific truth and its value for humanity are (1) an understanding of the psychological constitutions of the investigator and (2) an understanding of the social forces which produce him, encourage or oppose him, and transmit or ignore his work" (1938, p. 50).

Not surprisingly, the period surrounding WWII was characterized by increased European attention, at least, to the interrelation of science and values, to which an important contribution was made by Gestalt psychologist Wolfgang Köhler (1938) that articulates the formative relation of human values to any inquiry, drawing on principles gleaned through Gestalt research and theory. To this we should add that George Kelly's metaphor of "person as scientist" (1955), though not intended as a psychological analysis of science, implicitly affirms the personhood of the researcher and describes the nature of the thought processes that bind scientific thinking to everyday cognition and meaning-making. Skinner demonstrated interest in the psychology of science in various ways. For example, in an essay titled "Can Science Help?" he calls science "a unique intellectual process with remarkable results" (1953, p. 11), but "first of all a set of attitudes.…a disposition to deal with the facts rather than what someone has said about them" (1953, p. 12).

Elsewhere he briefly engaged Polanyi's arguments for the personal contribution of the scientist to the process of scientific knowledge production (Polanyi, 1974/1958), but concluded that although the scientist "must behave as an individual"… "he [sic] states facts or laws which make it possible for others to respond effectively without personal exposure to that world" (Skinner, 1974, pp. 144–145).

Feist and Gorman (2013) note that the psychology of science was more actively pursued in Eastern Europe during the postwar period. They highlight Mieczyslaw Choynowski's *Institute for the Science of Science* and its corresponding publication of a journal dedicated to describing its assumptions and findings (e.g., Choynowski, 1948), as well as a center established within the Russian Academy of Sciences focused on the psychology of scientific innovation and achievement. Feist and Gorman (2013) describe an uptake if not a surge of interest in the study of scientific creativity, including its cognitive organization and the personality characteristics that might predict it, in the decades of the 1950s through 1970s, of which studies by Roe (1953), Eiduson (1962), Taylor and Barron (1963), and Maslow (1966) are examples. Simonton (1988) credits the surge of interest in the early 1960s, at least on the part of American psychologists, to competition to win the space race, provoked by the launching of the Soviet shuttle. Although Simonton considered the psychology of science to be waning by the early 1970s, we should note an important exception in Ian Mitroff's qualitative (interview-based) analysis of NASA scientists, important not only for its contribution to the content of psychology of science but for its setting of a precedent for the contemporary strategy of interviewing scientists for the purposes of better understanding the nature of science (Mitroff, 1974). Simonton saw the momentum for psychology of science picking up again in the early 1980s, in keeping with the expansion of cognitive psychology as noted above. Building on the expanded study of creativity in the decades preceding, his own *Scientific Genius* offered a comprehensive new psychological theory ("chance configuration") to account for new combinatorial processes preceding or concurrent with creative insight (Simonton, 1988).

The first decade of the twenty-first century was witness to yet another renewal of interest in the psychology of science. Groups of researchers seemed to be operating independently, but Greg Feist attempted to organize their efforts toward the recognition of a distinct subdiscipline of the psychology of science. Feist (2006)'s book, titled *The Psychology of Science*, summarizes historical and contemporary contributions to the study of science and offers a theoretical and empirically informed account of the developmental origin of scientific thinking. His efforts to organize a subfield of psychology of science included an international conference and a series of workshops and published anthologies specifically dedicated to the psychology of science (e.g., Feist & Gorman, 2013; Proctor & Capaldi, 2012). These collections are skewed toward understanding the cognitive mechanisms underlying science—the mechanics of scientific reasoning. The publication of anthologies dedicated to psychology of science raises important questions about the boundaries of the subfield—what is included and what is left out: What is to be counted as the *psychological* study of science in historical and contemporary contexts?

In addition to the many lines of broadly psychological study already named, we must also recognize historical studies of our own discipline as occupying a space within the broad province of the psychology of science. Heidbreder's depictions of Wundt's "order-loving mind" dedicated to the preparation of "edition after edition of his formulated system" (1933, p. 96), or of the "emotional urgency with which the thoroughgoing behaviorist takes his beliefs and taboos" (p. 268) clearly illustrate integration of dispositional and motivational considerations into an account of scientific progress. Clearly the account is interpretive, not a controlled study, but it shines light on the science in a way that has helped generations of students comprehend differences in theoretical frameworks and their epistemic values. More contemporary histories are likely to emphasize the social, cultural, and political embeddedness of psychological theory, yet where would we draw a line enabling us to say that some aspect of this analysis is not psychological study? Similarly, critical analysis of psychological research and theory, whether aimed at the level of metacritique or focused analysis of particular psychological concepts or methods (e.g., Slaney, 2017; Tissaw, 2007) surely counts as psychological study of psychological science.

More to the point of this volume is that a range of theoretical and methodological standpoints contribute to the psychological study of science, though they have not been recognized as such. It is worthwhile to note what has been happening in philosophy of science, perhaps the most entrenched of the disciplines that seek to inform or illuminate scientific process. Dominated for decades and even defined by the methods of analytical philosophy, recent years have seen radical new approaches with data collection and analysis serving as the basis upon which to evaluate philosophical claims—an empirical philosophy of science (e.g., Ichikawa, 2012; Lombrozo, Knobe, & Nichols, 2014). Yet within the philosophical community there is widespread recognition that the inherent complexity of the underlying phenomenon (science) justifies and even requires a range of methods, and that the meaning of "empirical" is inclusive of qualitative and historical analysis (Osbeck & Nersessian, 2015). A new wave of fine-grained analysis of science "in the wild" puts taken for granted assumptions about the nature of science to the test and offers richer understanding of how knowledge construction is accomplished and propagated, and to what ends (Andersen, 2016; Leonelli, 2016a, 2016b). In particular, should a psychology of science be less inclined to recognize the contribution of qualitative and historical studies to the analysis of science as practiced? We assert that it should not.

## Psychological Studies of Technology

Sometimes in parallel with the study of science and sometimes integrated with it, academic study of technology has emerged as an important subfield. Again, we note a conspicuous absence: study of the psychological significance of the world of machines, systems, and techniques in human experience and action is largely absent from twentieth-century psychology. It is only within recent years that we see development toward a more systematic psychology of technology (Rosen, Cheever, & Carrier, 2015; Sørensen & Schraube, 2013; Turkle, 2008).

An important reason for the difficulty of psychology to contribute to the study of technology has to do with psychology's one-sided conception of itself as a science, primarily reliant on the foundation of a quantitative methodology. Quantitative, experimental, and statistical methodology decontextualizes psychological phenomena and separates them from the world, including the world of science and technology. Within such an approach the internal relationship between humans and technology and psychological meaning-making processes cannot really be addressed. Quantitative methodology disarticulates human subjectivity and undercuts the possibility of substantially investigating how human beings are involved in the making of the world of technology as well as what this world actually means for human experience and action.

Accordingly, STS scholars have argued for a systematic inclusion of human subjectivity in accounts of scientific research practice and technological production. On the basis of detailed empirical analyses of the production of scientific knowledge (Knorr-Cetina, 1981, 1999; Latour & Woolgar, 1986), STS scholarship calls for a notion of science and technology as socially situated processes that can only be adequately grasped through the study of their material, cultural, and subjective entangledness. For this reason, and from its earliest days, STS has questioned the idea of technological determinism, including mechanistic cause-and-effect understandings of the human–technology relationship. Instead, STS scholarship has based its scientific approach on a conception of human beings as subjects and collectives, subject to, but also actively involved in the production of the world of science and technology. As Langdon Winner emphasizes: "By changing the shape of material things … we also change ourselves. In this process human beings do not stand in the mercy of a great deterministic punch press that cranks out precisely tailored persons at a certain rate during a given historical period. Instead the situation … is one in which individuals are actively involved in the daily creation and recreation, production and reproduction of the world in which they live" (1989, pp. 14f.). STS acknowledges that the possibility of examining the nature and significance of technological artifacts in human experience, thought, and action requires a careful consideration of the subjective dimension of human life.

As Sherry Turkle explains: "Technology catalyzes changes not only in what we do but in how we think. It changes people's awareness of themselves, of one another, of their relationship with the world.... My focus is ... on the 'subjective computer'. This is the machine as it enters into social life and psychological development, the computer as it affects the way we think, especially the way we think about ourselves" (1984, pp. 18f.).

Moreover, throughout the history of psychology, we can find various traditions of thought seeking to include the subjective dimension of human life in their scientific vision and develop a psychological vocabulary capable of articulating in detail the complex relationship between individual subject and societal world. It is exactly within these approaches that we can find a *turn to technology* and sophisticated investigations of the psychological implications of technology in human life; for example in studies building on psychoanalysis, activity theory and sociocultural approaches (Schachtner, 2013; Turkle, 2015; Valsiner, 2014), social constructionist, discursive and post-structuralist approaches (Gergen, 2000; Gordo-López & Parker, 1999; O'Doherty & Einsiedel, 2013; Søndergaard, 2013), critical psychology (Costall & Dreier, 2006; Schraube, 2009, 2013; Schraube & Marvakis, 2016) or actor–network theory and posthumanist philosophy (Sørensen, 2009). The decisive step of these approaches is to engage in developing a psychological theory and methodology which facilitate the articulation of the human–world relationship including the world of science and technology. Especially since the early 1970s, a multiplicity of psychological traditions of thought emerged that approach psychological phenomena not only in the abstract form of numbers and variables, but as subjective processes and activities of persons (Blackman, Cromby, Hook, Papadopoulos, & Walkerdine, 2008). Theoretical concepts such as *self, identity, subjectivity* or *agency* became influential for exploring the subjective dimension of psychic life and persons' embodied experiences, perspectives, self-understandings and engagements as deeply embedded in, and shaped by social structures and political arrangements.

However, including subjective processes in the epistemic framework is not a guarantee for overcoming the separation between person and world in psychological theory and research practice. Focusing on subjectivity can lead to individualistic scientific visions with research centered solely on individual subjects and the I-perspective, without seeing psychological processes in their social, historical, and technoscientific contexts. This danger is addressed, for instance, by Papadopoulos (2003) and Parker (2014) and we can build on a body of sophisticated historical analyses of human subjectivity which reveal in-depth how psychological processes are never just subjective, but always also *contextual processes situated in the world* (Harré, 1979; Holzkamp, 2013; Valsiner, 2014; Vygotzky, 1978).

Including the subjective dimension of human life in our studies therefore means taking subjective phenomena as a starting point, but—because the subjective is rooted in the social, cultural, discursive, and technoscientific world—studying them in their worldly connections. In the words of Ole Dreier: "To gain a richer, more concrete and lively theoretical conception of the person, we must, paradoxically, not look directly into the person but into the world and grasp the person as a participant in that world" (2008, p. 40).

To refine the vocabulary of subjectivity and bring psychological theory and research practice closer to what it means and takes to be a subject living *in* the contemporary societal and technological world, some scholars have recently started to work with the concept *conduct of everyday life* (Dreier, 2016; Holzkamp, 2013, 2016; Schraube & Højholt, 2016). Conduct of everyday life refers to the activities of individual subjects in arranging and organizing their everyday living. The concept integrates various psychological functions and processes (such as experiencing, feeling, thinking, learning, acting) and situates these processes integrally in their relations to other persons and the everyday world. The concept involves tracing how people make sense of the multiplicity of socio-material relations and contradictory demands in and across the different contexts in which they are engaged in the common day-to-day; and it takes into account how persons collaboratively produce and reproduce their life through daily activities, habits, rhythms, and routines. Since the concept focuses attention on the ways in which persons *organize and arrange* their everyday living, and technological artifacts play an essential role in everyday life, questions concerning human–technology relations are integral to the concept.

How does this concept substantiate our theoretical understanding of the connection between person and technoscientific world? Based on the centrality of *meaning*, it suggests including *subjective reasons for action* as a second central mediating dimension of the relationship between the individual subject and the socio-material world. *Social and technological conditions* never appear to the subject as such, but are always mediated through *societal configurations of meaning*. The concept of meaning replaces the cause-effect and stimulus-response logic of variable psychology and is therefore a key concept in the psychological study of science and technology. Persons live their everyday lives within the context of interwoven and interrelated societal meaning constellations which articulate more or less adequately the meaning of the respective societal condition. However, the configurations of meaning do not just determine persons' everyday experience and activity (such a view would simply take us back to a causal model of explanation), but represent for the individual subject-specific possibilities for (and constraints on) action, which each person—depending on their *subjective reasons for action*—can, but by no means must, enact in his/her conduct of everyday life. As a social configuration of meaning, the respective conditions flow into the premises of each person's subjective reasons for action. By recourse to these premises, individuals perceive their own actions as grounded and evaluate how far the respective conditions reflect their own life interests, or their need to act individually or collectively to increase their influence over the societal conditions of their conduct of everyday life (Holzkamp, 2013, pp. 281ff.). In short, persons' conduct of everyday life is not simply affected or conditioned by technoscientific conditions, but grounded in them as possibilities for action. In this sense the concept *conduct of everyday life* can help refine the understanding of the connection between person and world, particularly by articulating the subjects' experiences, reasons and the scope of action as they are grappling with the technoscientific conditions of their everyday activities. In this way, a psychology that grounds human subjectivity and agency in their sociomaterial bases provides potentially fruitful conceptual resources for the study of human entanglements with technoscience.

## Psychological Contributions to STS

As we have discussed, there is a long history of inquiry that can be considered both distinctively "psychological" and also potentially useful for questions pursued in STS scholarship. It goes without saying that psychological research has informed some of the earliest sociohistorical inquiry of STS scholars. For example, Thomas Kuhn (1962), in *The Structure of Scientific Revolutions*, drew upon Bruner and Postman's (1949) experimental research on perception to develop his arguments about the theory-ladenness of observation. It is nevertheless the case that psychological perspectives per se are not well represented within STS. There are three possible reasons for this. First, the discipline of psychology has been, for the most part, unreflexive about its own epistemological commitments. Thus, while STS has worked to disrupt received views of science as an entirely formal, rational activity, psychology has clung to its identity as a science whose methods produce knowledge that transcends the sociohistorical conditions of its production. Psychological accounts of scientific thinking, personality dynamics of scientists, or the inheritance of creative traits might therefore be incompatible with the more situated accounts preferred in STS work. Secondly, psychological inquiry is often hamstrung by a sequestrated ontology that artificially divides human life and subjectivity into domains that reflect psychology's various subdisciplines or fields. Thus we have "biological psychology of science", "personality psychology of science", "social psychology of science", "discursive psychology of science", and so forth (see, for example, Feist, 2006). Psychology is also dogged by a set of modernist dualisms that make it difficult to adequately conceptualize or grasp technoscientific phenomena in their sociomaterial specificity. Given that some of the work central to STS scholarship fundamentally questions such ontological distinctions, it is perhaps easy to see why psychological research is taken up in mostly piecemeal fashion. Indeed, in the process of developing accounts of technoscience, some STS scholars have attempted to rework categories (e.g., the "social", "human", "nature", and "culture") that have been central to modern social and psychological sciences (e.g., Barad, 1996; Haraway, 1985; Latour, 2005). Furthermore, situating technoscience has necessitated, in STS, analytical accounts showing the concurrence between scientific discovery, technological innovation, and societal change, marking a continuity or dynamic interrelation between individual scientists, the sterile confines of the laboratory and the social world (see, for example, Latour, 1993a; Law & Mol, 1995). Third, where psychologists themselves have produced more historically and socially contextualized accounts of technoscience, these have either veered away from perspectives that might be identified as distinctively psychological, or reproduced some of the troublesome ontological issues mentioned above. Social constructionism, for example, has been taken up in very particular ways in psychology, in some cases inspiring analytic accounts that are entirely silent on such dimensions as subjectivity, affect, history, or materiality.

All of this begs the question of what forms of inquiry can or should be considered "psychological" and whether this is in fact a helpful question at all. In order to avoid reproducing problematic disciplinary and ontological distinctions our account of "psychological" contributions to STS has been deliberately broad and transdisciplinary. Furthermore, given the plurality of conceptual and theoretical perspectives evident in much of this work, both historically, and in this collection, it makes little sense to rigidly demarcate the boundaries of such work. Yet there are strands that tie together the work which we wish to highlight in this volume. These are: the attempt to analyse the ontology, epistemology, and investigative practices of psychology through historical, theoretical, and empirical inquiry; the rehabilitation or reconstruction of psychological theory and conceptual categories to better account for technoscientific processes; and a focus on *subjectivity* or *conduct of everyday life* as core concepts for thinking the "psychological"—that is, an analytic category for understanding the social, historical, and material dimensions of human experience.

## Outline and Summary of Chapters

The chapters in this collection are structured according to three parts. The first part, *Scoping a New Psychology of Science and Technology*, features chapters that each provide a vision for an aspect of psychological studies of science and technology that goes beyond current conceptions of the field. In this context, Thomas Teo (Chapter 2) introduces the value of *epistemic modesty*. Teo argues that recognizing that the current state of science is always embedded in history and recognition of personal limitations of oneself (as a scientist) should come with an emphasis on modesty in making truth claims. However, Teo argues, this has not been the case and what he calls epistemic grandiosity is a much more common characteristic of scientific practice. In Chapter 3, Niklas Chimirri and Ernst Schraube call for a focus on subjectivity in the study of human–technology relations. They argue that technology needs to be understood as playing an important role in constituting subjectivity and the experience of everyday life. As such, they consider the outlines of a psychological study of technology that is based on an embodied and situated view, and that sees technological artifacts as contradictory and political forms of everyday life. In Chapter 4, Henderikus Stam examines criticisms of neuroscientific developments in psychology. While generally in agreement with these criticisms, Stam argues that these criticisms miss important elements of how neuroscience is changing our understanding of human phenomena. Stam introduces the idea of *epistemological first aid* to guide an understanding of the role of neuroscience in psychological theorizing.

Part II of the book, *Applying Psychological Concepts to the Study of Science and Technology*, provides a series of empirical and theoretical studies in which psychological concepts or approaches are used to make sense of or reinterpret scientific knowledge and practices. In opening this part, Lisa Osbeck and Nancy Nersessian (Chapter 5) report on an ethnographic study of research labs in four different fields of bioengineering science. Among other things, their study seeks to characterize interdisciplinary "thinking in action", and demonstrates how an ethnography of scientists can help to inform classroom design and curriculum development. In Chapter 6, Nora Ruck, Alexandra Rutherford, Markus Brunner, and Katharina Hametner examine how a consideration of psychological processes might usefully expand existing concepts in STS. In particular, they consider how epistemological concepts developed by feminist and critical race theorists might be further developed by considering the role of psychological processes such as affective dissonance.

Michael Pettit (Chapter 7) then conducts a historical analysis of the concept of the social network. Beginning in the discipline of psychology, the concept shapes assumptions about individuals and the relations they inhabit. Pettit explains key concepts from social network analysis and demonstrates how this form of analysis can provide important insights about science as a social practice. In the final chapter in this part (Chapter 8), Amanda Jenkins, Shannon Cunningham, and Kieran C. O'Doherty consider some of the potential social and psychological consequences of the emerging field of human microbiome science. While recognizing the likely positive impact this research will have on human health, Jenkins, Cunningham, and O'Doherty ask what some of the unintended and unforeseen social, psychological, and ethical ramifications might be for patients and their families.

Part III of the book, *Critical Perspectives on Psychology as a Science*, collects chapters that connect concepts and frameworks from STS scholarship to those of psychology. In some instances, this involves applying concepts from STS to an examination of different aspects of psychology; in others, the project is to identify contributions that psychology can make to STS. Estrid Sørensen (Chapter 9) opens this part with an explicit focus on the question of what value insights from STS can bring to psychology. Beginning with a reflection on the origins of psychology, Sørensen argues that by privileging an objective, quantitative, and detached foundation for psychological knowledge, the discipline has limited its own relevance to the world. Sørensen suggests that key insights from STS may help psychology redefine its epistemological foundation to produce a more satisfactory and relevant psychology. In Chapter 10, Nathalie Lovasz and Joshua Clegg develop a critique of evidence-based practice (EBP) as a foundation for psychological practice. Their analysis focuses on the processes through which the American Psychological Association task force on EBP came to constitute and define "evidence" in very particular ways. They show that what counts as evidence, in this context, is not self-evident, but rather the outcome of the negotiation of epistemic, political, practical, and interpersonal considerations. In Chapter 11, Kathleen Slaney, Donna Tafreshi, and Charlie A. Wu tackle the question of whether there is philosophical reflexivity in psychological science, and whether there should be.

They note that while reflexivity is a prominent theme in STS scholarship, it is not much in evidence in most conventional psychological science (nor, for that matter, in past attempts to develop a psychology of science and technology). Sapphira Thorne and Peter Hegarty (Chapter 12) turn their attention to how cognitive psychology might make a contribution to STS scholarship. They examine whether cognitive psychology can contribute to critical psychological projects and consider specifically what the implications are for LGBTQ+ psychology. In Chapter 13, Katherine Hubbard and Natasha Bharj observe some striking similarities in the ways in which women and the Rorschach ink blot test have been treated historically in psychology. Hubbard and Bharj cast a critical gaze on the discipline from a feminist STS perspective and examine attributions of subjectivity and how these have been used to delegitimize certain people and forms of knowledge, and thereby legitimize others. Finally, Ines Langemeyer (Chapter 14) concludes the collection with a critical reading of the work of Haraway and Barad and its relevance to orienting scientific research to questions relating to power, justice, and responsibility.

Taken together, these fourteen chapters cover a wide range of theoretical positions, methods, and scholarly traditions. They are not intended to provide a unified vision of a psychology of science and technology. To the contrary, our aim is to illustrate and underscore the importance of a pluralism in psychological contributions to the studies of science and technology. Our aim is thus not so much to replace previous articulations of a psychology of science and technology with a new or different foundation, but rather to open up the field to a wider range of approaches.

## References

Adorno, T., & Horkheimer, M. (2002). *Dialectic of enlightenment.* Stanford, CA: Stanford University Press (Originally published 1944).

Andersen, H. (2016). Collaboration, interdisciplinarity, and the epistemology of contemporary science. *Studies in History and Philosophy of Science Part A, 56,* 1–10.

Bachelard, G. (2002). *The formation of the scientific mind: A contribution to a psychoanalysis of objective knowledge*. Manchester, UK: Clinamen Press.
Bacon, F. (1937). *Novum organum* (R. F. Jones, Ed.). New York: Odyssey Press (Originally published 1620).
Barad, K. (1996). Meeting the universe halfway: Realism and social constructivism without contradiction. In L. H. Nelson & J. Nelson (Eds.), *Feminism, science and the philosophy of science* (pp. 161–194). Dordrecht: Springer.
Blackman, L., Cromby, J., Hook, D., Papadopoulos, D., & Walkerdine, V. (2008). Creating subjectivities. *Subjectivity, 22*(1), 1–27.
Bloor, D. (1975). A philosophical approach to science. *Social Studies of Science, 5*(4), 507–517.
Brentano, F. (1995). *Psychology from an empirical standpoint* (A. Rancurello, D. B. Terrell, & L. McAlister, Trans.). London: Routledge (Originally published 1874).
Bruner, J. S., & Postman, L. (1949). On the perception of incongruity: A paradigm. *Journal of Personality, 18*(2), 206–223.
Burman, E. (2016). *Deconstructing developmental psychology*. London: Routledge.
Campbell, D. T. (1989). Fragments of the fragile history of psychological epistemology and theory of science. In B. Gholson, W. R. Shadish, Jr., R. A. Neimeyer, & A. C. Houts (Eds.), *Psychology of science: Contributions to metascience* (pp. 21–46). New York: Cambridge University Press.
Choynowski, M. (1948, September). Life of science. *Synthese, 6*(5–6), 248–251.
Costall, A., & Dreier, O. (Eds.). (2006). *Doing things with things: The design and use of everyday objects*. Aldershot: Ashgate.
Craik, K. (1943). *The nature of explanation*. Cambridge: Cambridge University Press.
Danziger, K. (1994). *Constructing the subject: Historical origins of psychological research*. Cambridge University Press.
Dewey, J. (1938). *Logic: The theory of inquiry*. New York: Henry Holt.
Douglas, H. (2009). *Science, policy, and the value-free ideal*. Pittsburgh: University of Pittsburgh Press.
Dreier, O. (2008). *Psychotherapy in everyday life*. Cambridge: Cambridge University Press.
Dreier, O. (2016). Conduct of everyday life: Implications for critical psychology. In E. Schraube & C. Højholt (Eds.), *Psychology and the conduct of everyday life* (pp. 15–33). London: Routledge.

Eagly, A. H. (1987). *John M. MacEachran memorial lecture series; 1985. Sex differences in social behavior: A social-role interpretation.* Hillsdale, NJ, US: Lawrence Erlbaum Associates.

Eiduson, B. T. (1962). *Scientists: Their psychological world.* Oxford, UK: Basic Books.

Fancher, R. E. (1983). Alphonse de Candolle, Francis Galton, and the early history of the nature-nurture controversy. *Journal of the History of the Behavioral Sciences, 19*(4), 341–352.

Fanon, F. (1967). *Black skin, white masks.* New York: Grove Press (Originally published 1952).

Fechner, G. (1860). *Elemente der Psychophysik.* Leipzig: Breitkopf & Härtel.

Feist, G. J. (2006). *The psychology of science and the origins of the scientific mind.* New Haven: Yale University Press.

Feist, G. J., & Gorman, M. E. (1998). The psychology of science: Review and integration of a nascent discipline. *Review of General Psychology, 2*(1), 3–47.

Feist, G. J., & Gorman, M. E. (Eds.). (2013). *Handbook of the psychology of science.* New York: Springer Publishing.

Feyerabend, P. (1975). *Against method.* London: Verso Books.

Foucault, M. (1966). *Les Mots et les Choses.* Paris: Éditions Gallimard.

Foucault, M. (1975). *Discipline and punish: The birth of the prison.* New York: Vintage Books.

Fox-Keller, E. (1985). *Reflections on gender and science.* New Haven, CT: Yale University Press.

Fuller, S. (2002). *Social epistemology.* Bloomington, IN: Indiana University Press.

Gentner, D., & Stephens, A. L. (1983). *Mental models.* Hillsdale, NJ: Lawrence Erlbaum Associates.

Gergen, K. J. (1973). Social psychology as history. *Journal of Personality and Social Psychology, 26*(2), 309–320.

Gergen, K. J. (2000). *The saturated self: Dilemmas of identity in contemporary life.* New York: Basic Books.

Giere, R. N. (Ed.). (1992). *Cognitive models of science* (Vol. 15). Minneapolis: University of Minnesota Press.

Gooding, D. (1985). In nature's school: Faraday as a natural philosopher. In D. Gooding & F. James (Eds.), *Faraday rediscovered* (pp. 105–135). London: Macmillan.

Gordo-López, Á. J., & Parker, I. (Eds.). (1999). *Cyberpsychology.* London: Routledge.

Gorman, M., & Carlson, W. (1990). Interpreting invention as a cognitive process: The case of Alexander Graham Bell, Thomas Edison, and the telephone. *Science, Technology and Human Values, 15,* 131–164.

Guthrie, R. V. (2004). *Even the rat was white: A historical view of psychology* (2nd ed.). Upper Saddle River, NJ: Pearson Education.

Haraway, D. (1985). A manifesto for cyborgs: Science, technology and socialist feminism in the 1980s. *Socialist Review, 80,* 65–108.

Harding, S. (1986). *The science question in feminism.* Ithaca, NY: Cornell University Press.

Harré, R. (1979). *Social being: A theory for social psychology.* Oxford: Blackwell.

Heidbreder, E. (1933). *Seven psychologies.* London, UK: Century/Random House UK.

Holzkamp, K. (2013). Psychology: Social self-understanding on the reasons for action in the conduct of everyday life. In E. Schraube & U. Osterkamp (Eds.), *Psychology from the standpoint of the subject: Selected writings of Klaus Holzkamp* (A. Boreham & U. Osterkamp, Trans.) (pp. 233–341). Basingstoke: Palgrave Macmillan.

Holzkamp, K. (2016). Conduct of everyday life as a basic concept of critical psychology. In E. Schraube & C. Højholt (Eds.), *Psychology and the conduct of everyday life* (pp. 65–98). London: Routledge.

Houts, A. C. (1989). Contributions of the psychology of science to metascience: A call for explorers. In B. Gholson, W. R. Shadish, Jr., R. A. Neimeyer, & A. C. Houts (Eds.), *Psychology of science: Contributions to metascience* (pp. 47–88). New York: Cambridge University Press.

Husserl, E. (1954). *The crisis of European sciences and transcendental phenomenology* (D. Carr, Trans.). Evanston: Northwestern University Press (Originally published 1936).

Ichikawa, J. J. (2012). Experimentalist pressure against traditional methodology. *Philosophical Psychology, 25*(5), 743–765.

James, W. (1987). Pragmatism. In *William James: Writings 1902–1910* (pp. 479–624). New York: Library of America (Originally published 1907).

Johnson-Laird, P. N. (1983). *Mental models: Towards a cognitive science of language, inference, and consciousness.* Cambridge, MA: Harvard University Press.

Kelly, G. A. (1955). *The psychology of personal constructs* (Vols. 1–2). New York: Norton.

Knorr-Cetina, K. (1981). *The manufacture of knowledge: An essay on the constructivist and contextual nature of science.* Oxford: Pergamon.

Knorr-Cetina, K. (1999). *Epistemic cultures: How the sciences make knowledge.* Cambridge, MA: Harvard University Press.

Koch, S., & Leary, D. E. (Eds.). (1992). *A century of psychology as science.* Washington, DC: American Psychological Association.

Köhler, W. (1938). *The place of value in a world of facts.* Oxford, UK: Liveright.

Kuhn, T. (1962). *The structure of scientific revolutions.* Chicago: University of Chicago Press.

Latour, B. (1987). *Science in action: How to follow scientists and engineers through society.* Cambridge: Harvard University Press.

Latour, B. (1993a). *The pasteurisation of France.* Cambridge: Harvard University Press.

Latour, B. (1993b). *We have never been modern.* Cambridge: Harvard University Press.

Latour, B. (2005). *Reassembling the social: An introduction to actor-network theory.* New York: Oxford University Press.

Latour, B., & Woolgar, S. (1986). *Laboratory life: The construction of scientific facts.* Princeton, NJ.: Princeton University Press.

Law, J., & Mol, A. (1995). Notes on materiality and sociality. *The Sociological Review, 43*(2), 274–294.

Leahey, T. H. (2017). *A history of psychology: From antiquity to modernity.* New York: Routledge.

Leonelli, S. (2016a). *Data centric biology: A philosophical study.* Chicago: University of Chicago Press.

Leonelli, S. (2016b). Locating ethics in data science: Responsibility and accountability in global and distributed knowledge production systems. *Philosophical Transactions of the Royal Society A: Mathematical, Physical and Engineering Sciences, 374*(2083), 20160122.

Lombrozo, T., Knobe, J., & Nichols, S. (Eds.). (2014). *Oxford studies in experimental philosophy* (Vol. 1). Oxford: Oxford University Press.

Mach, E. (1896). *Contributions to the analysis of the sensations* (C. M. Williams, Trans.). Chicago: Open Court.

Malone, K., & Barbarino, G. (2009). Narrations of race in STEM settings: Identity formation and its discontents. *Science Education, 93*(3), 48–510.

Maslow, A. (1966). *The psychology of science: A reconnaissance.* New York: Harper & Row.

Mill, J. S. (1843). *A system of logic, ratiocinative and inductive: Being a connected view of the principles of evidence and the methods of scientific investigation* (Vol. 1). London: John W. Parker.

Mitroff, I. (1974). *The subjective side of science: Philosophical inquiry into the psychology of the Apollo moon scientists.* Amsterdam: Elsevier.

Morawski, J. G. (Ed.). (1988). *The rise of experimentation in American psychology.* New Haven, CT, US: Yale University Press.

Nersessian, N. J. (1984). *Faraday to Einstein: Constructing meaning in scientific theories*. Dordrecht: Martinus Nijhoff/Kluwer.
Nersessian, N. J. (1992). How do scientists think? Capturing the dynamics of conceptual change in science. In R. Giere (Ed.), *Cognitive models of science* (pp. 3–44). Minneapolis: University of Minnesota Press.
O'Doherty, K., & Einsiedel, E. (Eds.). (2013). *Public engagement and emerging technologies*. Vancouver: University of British Columbia Press.
Osbeck, L., & Nersessian, N. (2015). Prolegomena to an empirical philosophy of science. In S. Wagenknecht, N. J. Nersessian, & H. Andersen (Eds.), *Empirical philosophy of science: Introducing qualitative methods into philosophy of science* (pp. 13–35). Cham: Springer International.
Papadopoulos, D. (2003). The ordinary superstition of subjectivity: Liberalism and technostructural violence. *Theory & Psychology, 13*(1), 73–93.
Parker, I. (2014). Managing neo-liberalism and the strong state in higher education: Psychology today. *Qualitative Research in Psychology, 11*(3), 250–264.
Peirce, C. S. (1878). How to make our ideas clear. *Popular Science Monthly, 12*(January), 286–302.
Pickren, W., & Rutherford, A. (2010). *A history of modern psychology in context*. Hoboken, NJ: Wiley.
Polanyi, M. (1964). *Science, faith, and society*. Chicago: University of Chicago Press (Originally published 1946).
Polanyi, M. (1974). *Personal knowledge: Towards a post-critical philosophy*. Chicago: University of Chicago Press (Originally published 1958).
Popper, K. (2002). *The logic of scientific discovery*. London: Routledge (Originally published 1959).
Proctor, R. W., & Capaldi, E. J. (Eds.). (2012). *Psychology of science: Implicit and explicit processes*. New York: Oxford University Press.
Reichenbach, H. (1938). *Experience and prediction: An analysis of the foundations and the structure of knowledge*. Chicago: University of Chicago Press.
Robinson, D. N. (1995). *An intellectual history of psychology*. Madison: University of Wisconsin Press.
Roe, A. (1953). A psychological study of eminent psychologists and anthropologists, and a comparison with biological and physical scientists. *Psychological Monographs: General and Applied, 67*(2), 1–55.
Rorty, R. (1979). *Philosophy and the mirror of nature*. Princeton, NJ: Princeton University Press.
Rosen, L. D., Cheever, N. A., & Carrier, L. M. (2015). *The Wiley handbook of psychology, technology, and society*. Oxford: Wiley Blackwell.

Schachtner, C. (2013). Digital media evoking interactive games in virtual space. *Subjectivity, 6*(1), 33–54.

Schaffer, S. (1988). Astronomers mark time: Discipline and the personal equation. *Science in Context, 2*(1), 115–145.

Schraube, E. (2009). Technology as materialized action and its ambivalences. *Theory & Psychology, 19*(2), 296–312. https://doi.org/10.1177/0959354309103543.

Schraube, E. (2013). First-person perspective and sociomaterial decentering: Studying technology from the standpoint of the subject. *Subjectivity, 6*(1), 12–32. https://doi.org/10.1057/sub.2012.28.

Schraube, E., & Højholt, C. (Eds.). (2016). *Psychology and the conduct of everyday life*. London: Routledge.

Schraube, E., & Marvakis, A. (2016). Frozen fluidity: Digital technologies and the transformation of students learning and conduct of everyday life. In E. Schraube & C. Højholt (Eds.), *Psychology and the conduct of everyday life* (pp. 205–225). London: Routledge.

Simon, H. (1966). The psychology of scientific problem solving. In R. Tweney, M. Doherty, & C. Mynatt (Eds.), *On scientific thinking* (pp. 48–54). New York: Columbia University Press.

Simonton, D. K. (1988). *Scientific genius: A psychology of science*. Cambridge: Cambridge University Press.

Sinclair, B. (2004). *Technology and the African-American experience*. Cambridge, MA: MIT Press.

Skinner, B. F. (1953). *Science and human behavior*. New York: Macmillan.

Skinner, B. F. (1974). *About behaviorism*. New York: Alfred Knopf.

Slaney, K. (2017). *Validating psychological constructs: Historical, philosophical, and practical dimensions*. New York: Springer.

Søndergaard, D. M. (2013). Virtual materiality, potentiality and subjectivity: How do we conceptualize real-virtual interaction embodied and enacted in computer gaming, imagery and night dreams? *Subjectivity, 6*(1), 55–78.

Sørensen, E. (2009). *The materiality of learning*. Cambridge: Cambridge University Press.

Sørensen, E., & Schraube, E. (Eds.). (2013). Special issue: "Materiality". *Subjectivity, 6*(1), 1–129.

Taylor, C. W., & Barron, F. (Eds.). (1963). *Scientific creativity: Its recognition and development*. Oxford, UK: Wiley.

Teo, T. (2015). Critical psychology: A geography of intellectual engagement and resistance. *American Psychologist, 70*(3), 243.

Teo, T. (2017). From psychological science to the psychological humanities: Building a general theory of subjectivity. *Review of General Psychology, 21*(4), 281–291.

Tissaw, M. A. (2007). Making sense of neonatal imitation. *Theory & Psychology, 17*(2), 217–242.

Turkle, S. (1984). *The second self: Computers and the human spirit*. New York: Simon & Schuster.

Turkle, S. (Ed.). (2008). *The inner history of devices*. Cambridge, MA: MIT Press.

Turkle, S. (2015). *Reclaiming conversation: The power of talk in a digital age*. New York: Penguin Press.

Tweney, R. D. (1989). A framework for the cognitive psychology of science. In B. Gholson, W. R. Shadish Jr., R. A. Neimeyer, & A. C. Houts (Eds.), *Psychology of science: Contributions to metascience* (pp. 342–366). New York: Cambridge University Press.

Tweney, R. D., Doherty, M. E., & Mynatt, C. R. (Eds.). (1981). *On scientific thinking*. New York: Columbia University Press.

Valsiner, J. (2014). *An invitation to cultural psychology*. London: Sage.

von Helmholtz, H. (1995). The recent progress of the theory of vision. In D. Cahan (Ed.), *Science and culture: Popular and philosophical essays* (pp. 127–203). Chicago: The University of Chicago Press (Originally published 1868).

Vygotzky, L. S. (1978). *Mind in society*. Cambridge, MA: Harvard University Press.

Wagenknecht, S., Nersessian, N. J., & Andersen, H. (Eds.). (2015). *Empirical philosophy of science: Introducing qualitative methods into philosophy of science* (vol. 2). Springer. Studies in Applied Philosophy, Epistemology and Rational Ethics. https://doi.org/10.1007/978-3-319-18600-9.

Watson, D. L. (1938). *Scientists are human*. London: Watts & Co.

Wertheimer, M. (1981). Einstein: The thinking that led to the theory of relativity. In R. Tweney, M. Doherty, & C. Mynatt (Eds.), *On scientific thinking* (pp. 192–211). New York: Columbia University Press.

Whewell, W. (1847). *The philosophy of the inductive sciences* (2 Vols.). London: John W. Parker.

Winner, L. (1989). *The whale and the reactor: A search for limits in an age of high technology*. Chicago: University of Chicago Press.

Wittgenstein, L. (1953). *Philosophical investigations* (G. E. M. Anscombe & R. Rhees, Eds., G. E. M. Anscombe, Trans.). Oxford: Blackwell.

# Part I

## Scoping a New Psychology of Science and Technology

# 2

# Academic Subjectivity, Idols, and the Vicissitudes of Virtues in Science: Epistemic Modesty Versus Epistemic Grandiosity

Thomas Teo

Epistemic subjectivity has been the nemesis of objectivity. To be more precise: Subjectivity has always been a dialectical part of objectivity (see Daston & Galison, 2007). Objectivity is not only an epistemic category, but also a *value* that guides science (Teo, 2018a). To demand from the subject to "be objective" is clearly a normative claim and shows the connection between epistemology and ethics, or, what one could label "epistemo-ethics." Epistemic values become personal virtues once they are considered positive and embodied in concrete subjectivities. Scientists, implicitly and explicitly, have committed to various epistemic virtues over time. Traditional values may include academic freedom, honesty, transparency, truth, or objectivity, while critical researchers may emphasize truthfulness, and social, economic, and environmental justice as ideals of research.

Whereas objectivity remains a widely endorsed value and virtue, *epistemic modesty* is hardly mentioned in textbooks, the academic literature,

T. Teo (✉)
Department of Psychology, York University, Toronto, ON, Canada
e-mail: tteo@yorku.ca



K. C. O'Doherty et al. (eds.), *Psychological Studies of Science and Technology*, Palgrave Studies in the Theory and History of Psychology,
https://doi.org/10.1007/978-3-030-25308-0_2

or seminars, although it is a value that emerges when understanding subjectivity in its full complexity, including a recognition of the historical, cultural, and personal limitations of knowledge. While the historicity and the personal limitations of knowledge apply to all sciences, culture-centric knowledge within a complex globalized world impacts the human sciences more deeply. Yet, it also leads human scientists to be anxious about the consequences of humble knowledge claims in the public domain. Epistemic subjectivity may also be compared to everyday subjectivity and in this process may be aligned with *epistemic grandiosity*. The dialectics between epistemic modesty and grandiosity in human subjectivity has a long history in European thinking.

In Western philosophy this conflict is played out in classical Graeco-Roman expositions. Cicero's (106–43 BCE) Socrates laid the foundation for understanding the limits of one's knowledge, for being open to the uncertainty of one's knowledge, translated from Greek, and later from Latin, as the dictum: "I know that I do not know" (sometimes erroneously translated as "I know that I know nothing") (see Fine, 2008). This must be contrasted with Plato's (1997) allegory of the cave, involving Socrates as well, that portrays the knower as belonging to the few who embrace truth, against the many who chain themselves to ignorance and who sacrifice the true knower in a seemingly inevitable course of events. Western philosophy has contributed to this mindset in the works of Immanuel Kant (1724–1804), who labeled his own contributions a Copernican revolution, or Georg W. F. Hegel (1770–1831), who suggested that the absolute spirit was embodied in his works, or Friedrich Nietzsche (1844–1900), who titled one of his chapters in *Ecce Homo* (written 1888), arguably ironically, "Why I write such good books," to mention only a few examples.

If we put Francis Bacon (1561–1626) at the formation of modern Western science, we find a similar dialectic (one could include Descartes): His idols emphasize the importance of understanding the hindrances to knowledge (Bacon, 1965), whereas the limitations do not apply to himself, and his own statements lack modesty (see also Keller, 1985). It is a common current in scientific thought that the limitations of knowledge, a lack of objectivity or rigor, or incompetence are attributed to other people but not to oneself. Accusing other researchers of bias, ignorance, and speculation has been a tool to diminish the epistemic quality of the work of others (Teo, 2008). This applies equally to "positivist" and empiricist approaches, as well as to critical scholarship that should be aware of its own temporality.

The belief that one can assume a point from nowhere, that history, culture, and society do not play a role in epistemic subjectivity, that "I" am objective, whereas others are not, may lead to a feeling of epistemic grandiosity, whereas the assumption that "my" knowledge is always fragile, even when "I" attempt to be objective, might inspire epistemic modesty. In the natural sciences, modesty could include the context of discovery (see Reichenbach, 1938) (what questions are asked and why), and one's own inevitably narrow expertise, whereas in the human sciences, it applies to the contexts of discovery, justification (how was a claim justified? what methodology or method was used to make this statement?), interpretation (how were results interpreted?), and application (how were findings translated for practical purposes?). Particularly in the human sciences, the temporality and contextuality of objects and events demands making epistemic modesty a virtue.

For instance, postcolonial research has shown the degree to which Western ideas permeate knowledge in the human, social, and psychological sciences. Such research identifies power dynamics against the *periphery* that include misrepresentations, silencing, and structural and epistemic violence (e.g., Spivak, 1999). Scientific projects have played an important role in *othering* non-Western mental life (Jackson & Weidman, 2004). Western science has shown the cultural, colonial, and indigenous impact of knowledge, and how colonial interests have been responsible for the generation of knowledge about dominated people. Even when there is an agreement that European history is not world-history, historians from the "periphery" (e.g., India), still need to address the "center's" history, which does not hold true for the center (Chakrabarty, 2000). Similarly, human scientists from the periphery must relate indigenous knowledge to mainstream organizations, journals, and practices if they want to have an impact, whereas the opposite is not required of the mainstream.

Indigenous knowledge has shown that we can have alternative conceptualizations about the human world (e.g., Kim, Yang, & Hwang,

2006), and that much of Western thinking itself is indigenous in its context. The problem is not just one of sampling or organizational structure (Henrich, Heine, & Norenzayan, 2010), but affects the core categories with which disciplines in the human sciences operate. The problem of *epistemic ethnocentrism* is not confined to the participants of research, but includes distortions and interests that emerge from hypotheses, interpretations, and research practices that psychologists have adopted. Power and culture play a role in the choice of problems, methods, data analyses, discussions, and applications. The solution to the problem of ethnocentrism in the human sciences is not about expanding but about decolonizing such sciences (see Adams & Estrada-Villalta, 2017; Bhatia, 2018). Psychological intuitions, categories, theories, philosophies, and even methodologies have a cultural dimension embedded in power. Thus, for subjects involved in sciences about humans, the value of modesty in regard to one's knowledge claims should be obvious. Epistemic virtues and values—endorsed or embodied—are one area of research for a new psychology of science.

## A Psychology of the Sciences and Beyond

Attempts to capture some of the subjective elements in the scientific process have been accomplished in a traditional psychology of science. For instance, one of the pioneers of psychology, Galton (1869/1962), proclaimed that scientific genius was inherited, as well as that modern Europeans are in greater possession of it than other races. The botanist Candolle (1873), skeptical of Galton's nature over nurture arguments, agreed that colored races lack men of scientific discovery and that women have not written any original scientific work. The criminologist Lombroso (1905) believed that the giants of the mind may be burdened with mental illness, while the Nobel Prize recipient and chemist Ostwald (1908) heightened the appreciation of scientists in suggesting that scientific innovators are the most important class of human beings, contributing to epistemic and social grandiosity and not to modesty.

In a largely forgotten, systematic book on the psychology of science, Hiebsch (1977) suggested that the subdiscipline studies creative thinking, the ways in which the personality of the problem-solving individual conditions cognitive activity, and how creative thinking and the creation of knowledge can be advanced through working in teams. For instance, Wertheimer (1945) analyzed productive thinking by looking at the thought processes of Galileo Galilei (1564–1642) and Albert Einstein (1879–1955), and how these thought processes led to the beginning of modern physics and the development of the theory of relativity. Guilford (1950) in his *Presidential Address* expressed the importance of the discovery and development of creative talent, since creativity by scientists and engineers had economic value. Yet, attempts to improve the practice of science have not involved psychologists, but authorities in business and management.

The case for a psychology of science was made by Müller-Freienfels (1936), who argued that epistemologies produced abstract systems that ignore living human beings that produce knowledge. In contrast, the philosopher of science Popper (1972) famously banned the psychology of science from epistemology and pleaded for an epistemology without a knowing subject. Yet, Fleck (1935/1979) incorporated with his concepts of *thought style* and *thought collective* a social psychology into his understanding of science. Kuhn (1962) can be understood as including within a psychology of science the idea that the inability to find a solution challenges the researcher and not the theory, that students accept theories because of the authority of a teacher, or that most scientists perform normal science.

Psychological ideas as core to the study of science have been endorsed by historians of science. For instance, Holton (1973) moved the argument into the direction of psychology by arguing that thematic decisions by individuals are more important than paradigms, and that such commitments emerge from the personality of the individual, rather than from the environment or community of the researcher. A *thema*, just like in music, is something that may be repeated and may recur throughout a scientific career. Even Feyerabend (1975) employed psychological insights in his anarchistic epistemology, by pointing to the obedience of researchers and the role of money and emotional support in the work of scientists. The question regarding the age at which scientists reach their peak is of psychological interest as well. Albert Einstein,

Werner Heisenberg (1901–1976), or Paul Dirac (1902–1984) experienced their peak before the age of 30, and it appears that physicists at that time lost their creativity after the age of 35; chemists reached their peak at 40; and philosophers can improve into their 60s (Oeser, 1988). More recently, the works of Simonton (1988) or Feist (2006) fall under the umbrella of traditional psychology of science.

My critique suggests that most of the traditional psychology of science works do not put subjectivity in its nexus at the center. Even some critically oriented approaches ignore that nexus. Freud (1977) argued that the sublimation of sexual desires is also responsible for the highest cultural, artistic, and social achievements of humanity. More challenging for the mainstream is Devereux (1967), who connected anxiety and methods in the human sciences, and argued that the experiment in psychology is as much an experiment on the experimenter as it is on the participant. The anxieties and defense mechanisms of the researcher, research strategies, and the collection and interpretation of data disclose more about the nature of human behavior than the seemingly objective observation of rats or other human beings. There exists also an important tradition in feminist philosophy of science that points to the gendered psychological dimensions of science. The choice for quantification, the analyses of variables, and the preference for abstract conceptualizations may represent a masculine attitude toward problems (Code, 1993; Harding, 1986; Keller, 1985).

A critical psychology of science that embraces subjectivity needs a theory that encompasses the social (socio-subjectivity: culture, society, history, etc.), the interpersonal (inter-subjectivity: groups, peers, organizations, teachers, etc.), and personal dimensions (intra-subjectivity: mind and embodied practices, thinking, feeling, and motivation) in their nexus and in connection with the material worlds. In short, epistemic subjectivity requires the first-person standpoint of researchers in its interconnection with social reality (see also Schraube, 2013; Teo, 2017). For instance, a psychology of science needs to explain why researchers, using scientific methods and standards, have endorsed ideas and "knowledge" that turned out to be false and even violent (e.g., scientific racism, sexism, classism). Is this ideological knowledge the result of personality, cognitive mistakes, or group dynamics? If ideological knowledge were just a matter of personality or cognition, then it would be easy to overturn or combat it. But this type of knowledge has a long shelf-life because it represents historically constituted prejudices that have been materialized in social practices and then corroborated by the existing scheme of hypotheses testing (e.g., group differences; see Teo, 2008).

I submit that the exclusive focus on traditional psychological topics in research prevents an understanding of ideology, hidden assumptions, and taken-for-granted theories, and requires a theoretical shift from personality to a historical, cultural, and societal concept of subjectivity, a concept that includes interaction with peers and colleagues, but does not neglect unique personal characteristics. Such a concept also includes an analysis of the scientific habitus (Bourdieu, 1988), the embodiment of scientific activities, and the privilege to speak on behalf of truth (see Teo, 2016). Research in the human sciences that fortifies existing privileges cannot be sufficiently understood by focusing on the individual, cognition, or even groups. The concept of subjectivity (in its broad meaning) is able to more adequately capture the problem, and its anchoring in the psychological humanities allows for critical analyses that include the socio-and historical constitution of mental life, while not neglecting individual commitments and idiosyncrasies. Changing epistemic virtues and values express a culture *and* an individual commitment.

## New Idols

The statement "the more one knows, the more one knows what one does not know" is a play on Socrates and should have consequences for a study of epistemic values. There is an inherent conflict between epistemic modesty, engendered by such sayings, and the need to present oneself as part of the epistemic elite or as one of the grandiose minds of knowledge (especially when one has expert knowledge). The point is not that the system of knowledge deserves no recognition, because it does. Rather, the reality is that the individual scientist can accommodate only a tiny part of any system of knowledge, even within one discipline such as psychology. Scientists are not exempt from cognitive issues and emotional attachments (see also Osbeck, Nersessian, Malone, & Newstetter, 2011). Like all humans, they must deal with realities that emerge in the cultural intersection of socio-, inter-, and intra-subjectivity. In the following, the new idols of research, and the consequences of a neoliberal academia and a post-truth society, as they impact epistemic subjectivity and modesty, are discussed.

Epistemic grandiosity encourages and is nourished by various idols that make epistemic modesty a difficult proposition. The first set could be called (a) "idols of the narcissistic halo," for which celebrities are known (and are used in advertising), but which also touches academic subjectivity. This concept refers to the tendency for scientists who are recognized experts in one area to appear or present themselves as competent in other knowledge areas as well. Scientists afflicted with this condition bank on their accumulated cognitive capital to convince the public and other audiences about their all-around knowledge capabilities. This tendency may be nourished by the status of scientific methods, which, certainly, do not make the scientist an automatic expert in all knowledge content areas. Doubt and critique, modesty and humility, are abandoned for one's own thoughts and statements, partial knowledge, rhetoric and exaggerations, which in turn reinforce notoriety in the public. As examples, natural scientists and economists come to mind who claim to possess expertise about gender differences; or, one can consider William Shockley (1910–1989), the Nobel prize winner in physics in 1956, and his epistemic support for a political, scientific racism (see Tucker, 1994).

The second set, (b) "idols of ideology," refers to a process where experts in one area do not challenge their assumptions as experts but rather justify the status quo by providing scientific discourses. In the context of scientific racism, an example would be an expert who misunderstands the history of racism, and the power of the interpretation of results, while neglecting disconfirming evidence. In political economy, where this phenomenon was first observed (Marx, 1867/1962), this refers to the propagation of the sources of the constitution of wealth and the degree to which one's theoretical preference is embedded within one's own interest or the interest of the powerful (or see Graeber, 2011, for a more recent example about the origins of debt). This subjectivity involves a lack of awareness of the assumptions or underlying motives that lead to the promotion of certain knowledges that are used to justify the status quo as natural and inevitable. Although a description can be wrong, it is presented as normative. A recent political example from psychology is the defense of torture, when psychologists' presence in enhanced interrogations was interpreted as the absence of torture per se and provided the justification for the continuation of enhanced interrogations (Aalbers & Teo, 2017).

The third set, (c) "idols of bullshit" (see Frankfurt, 1986/2005), is exemplified by scientists and psychologists in the service of the tobacco industry. These scientists (including psychologists) were *bullshitters,* in the sense that arguments such as "correlation does not mean causation," "nothing has ever been proven definitively," "we have to understand the times," and so on, misrepresented what actually went on without being false (see also Oreskes & Conway, 2010). The argument that large-scale epidemiological studies do not demonstrate causality in a psychological sense is correct, but it assumes that a psychological understanding of causality (which is different from the understanding of causality in physics) is superior to the one in epidemiology. Another example of this idol would be the concept of heritability. Scientists often suggest, erroneously, that this concept denotes the degree to which an *individual* has inherited a trait, when in reality it is a *population* statistic. Bullshiters exaggerate, they present something local as being true around the world, and they provide misleading statements that appeal to a parochial common sense, all while knowing that they are not doing justice to the complexity of the problem. They pick and choose, ignore disconfirming evidence, take things out of context deliberately, and do all this with a sense of epistemic grandiosity.

Sometimes the idols of bullshit cannot be distinguished from the fourth set, (d) "idols of ignorance," especially when the bullshitter starts to believe that what they are promoting is true, and when bullshit morphs into "truth." However, this process is neither apolitical nor benign. The production of ignorance is sociopolitical and benefits existing power structures and economic interests. Proctor and Schiebinger (2008) focus on the cultural production of ignorance with scientists having a part in this production. Psychology, for instance, produces ignorance when focusing on the individual and excluding social conditions. The idea that all change begins with individuals, or the focus on individuals, ignores research on inequality that identifies the many negative consequences of inequality for the mental health of individuals (Wilkinson & Pickett, 2009). If mental health issues are embedded in inequality, which is a social and structural category, not a psychological category, then it is ignorance-producing to suggest that one can solve mental health issues on an individual, psychological level. Of course, this finding requires modesty as well.

## The Vicissitude of Epistemic Modesty

We are all ignorant on certain issues at various times and the personal limitations of knowledge should logically lead to modesty and not to grandiosity. Even if we try to overcome personal knowledge deficiencies, our knowledge deficit will always be larger than our knowledge surplus. Epistemic modesty is the consequence of acknowledging subjectivity, culture, history, and society in knowledge-making and -dissemination, especially in the human sciences. Some of the same figures that lacked humility also advocated for modesty (see also Grenberg, 2005). In my argument, epistemic modesty means to be aware of one's *own* horizon, the strengths and limitations of one's own approach, while being knowledgeable about the history, sociality, and culturality of knowledge.

Modesty means having an awareness of one's own accomplishments without assuming the superiority of one's own knowledge or taking on a paternalistic attitude toward the other. Modesty, which is based on self-understanding and self-respect (Grenberg, 2005), means being careful about the old and new idols of research. Modesty does not imply relativism, that anything goes, or that one is weak, inferior, self-degrading or self-contemptuous. Modesty does not mean rejecting one's own knowledge competencies. Rather, modesty refers to a realistic assessment of the possibilities and limitations of "my" knowledge, while neither overestimating nor underestimating these possibilities or limitations. Epistemic modesty is a historical outcome of all the research that has accumulated over the centuries, in different countries, with different groups. In psychological practice, it has been acknowledged that an epistemologically humble clinical approach may be better received by patients (Fowers, 2005; Hersch, 2006).

The question emerges as to why scientists have not developed more epistemic modesty. Why do many scientists prefer grandiosity? It should be clear, based on my short description of subjectivity, that virtues or idols cannot be understood without the larger context, and without moving from the internal to the external logic of research. There are reasons why epistemic grandiosity thrives and epistemic modesty starves. The analysis of an epistemic virtue as being endorsed by an individual or as the result of a philosophical argument is insufficient, if one does not take the larger societal context into account. There are at least two important societal factors that counteract epistemic modesty: the neoliberal transformation of societies in recent decades and the emergence of a post-truth cultural reality.

The neoliberal transformation of society involves the privatization, individualization, and marketization of common goods (Harvey, 2005), with enormous consequences for the subject's conduct of everyday life. Neoliberalism denotes materially a political-economic alteration that has taken place since the 1980s, and ideologically to a thought system that emphasizes the self and family in the market place to the degree that a *homo neoliberalus* has emerged (Teo, 2018b). At the psychological level, neoliberalism means the psychologization, responsibilization, and subjectification of persons. At the institutional level, it means that all public entities are affected by a transformation, universities and colleges included, that demands that they be managed like businesses.

A neoliberal academia (see e.g., Smyth, 2017) means that unsuccessful departments or programs are closed, whereby success is defined by financial outcomes and not by the quality of work. The number of administrators is increased to supervise faculty, and likewise the number of performance evaluations are increased for everyone in the name of accountability. A neoliberal academia entails for-profit calculations, academic output that is compared and ranked in contrast to other universities, salaries based on external performance criteria, and the devaluation of academic service work that does not involve revenue, profit, or other financial gains. At the same time, the same service work by administrators is lauded as managerial. Entrepreneurship is celebrated, precarious work for students and part-time faculty is more prevalent, and critically oriented humanities are increasingly devalued if they do not produce something that can be sold.

What does a neoliberal university do to academic subjectivity, and what values are thereby promoted? Certainly not promoted is the value of epistemic modesty, which conflicts in significant ways with a neoliberal academia, where advertising, selling, and the impact of one's research are measured and used as benchmarks for status, success, and promotion. Academic subjectivity needs to exaggerate and focus on the impact and promotion of one's research. Epistemic modesty is superseded by an entrepreneurial self that needs to look constantly at citations and impact. One can even ask to what degree fraudulent work in academia can be understood on the background of neoliberalism. Faculty need to market their research, and if not marketable, move to new products that promise grants and higher impact. The work of the world-renown expert on medieval history (with a limited number of citations due to the small community) does not count as much as the normal science of the neuroscientist who produces an average number of citations in their field. Incommensurability of research has been reduced to quantifiable measures. In short, neoliberal academia does not provide forms for the embodiment of modesty, and rather promotes an innovative self that is in the business of marketing all accomplishments, to the point where epistemic grandiosity appears as a natural outcome.

The second cultural context that counteracts epistemic modesty is the post-truth society (see McIntyre, 2018). Intellectuals who share a skepticism toward a *Truth* concept with a capital *T*—a stream of argument one can find not only in postmodern theory, but also in German Idealism and in Popper's (1935/1992) critical rationalism,—find themselves forced to defend the practice of science, the concepts of truth and evidence, as well as better and worse knowledge, all within a context where truth has lost its meaning, and opinions and feelings have the same status as careful knowledge and well-developed, systematic thought. In the public domain, this means defending scientific truth as a benchmark against which other claims can be measured. Although the scientist understands the degree to which knowledge is provisional, the public demands authoritative statements, especially when the opponents of scientific truth promote their claims in absolute terms. Epistemic schemes that require the public to buy-in, and for emotional or financial reasons are distributed widely, cannot be overcome through reason, especially when a scheme seems to provide tangible emotional advantages or privileges.

In this context, the modest knower, defending distinctions between better and worse interpretations, applications, and knowledge, and perhaps even relevant and irrelevant questions, will always lose against the apodictic claimant who announces truth with grandiosity. Although epistemic modesty emerges from the logic of research, a post-truth cultural reality urges academics to advocate for the authority of science and the grandiosity of scientists (which is confused with the possible grandeur of science), which itself is a move that some of the critically oriented sciences must problematize as a political move. The selling of science, itself a new value, needs to learn from entrepreneurship, and modesty or moral generalizability are not the foremost concerns of business (see also Horkheimer & Adorno, 1947/1982).

Sometimes the movement against modesty is supported by unique disciplinary constellations. For example, in psychology, the low disciplinary ranking, the lack of a clear foundation, the fear of not being recognized or taken seriously as a real science, and the clash and confusion with pop-psychology, have all led to an inferiority complex (if one were to remain in the language of psychology). Moreover, such developments have made it difficult for psychology, or psychologists, to promote epistemic modesty. Historically and empirically, we find that psychologists have needed to exaggerate the scientific status of psychology, its comparability to physics and other STEM sciences, and its knowledge claims as a discipline and practice (Teo, 2018a). Within such a backdrop, any call for modesty will likely fall on deaf ears.

In contrast, the argument I am putting forth here is that a psychology of science needs to include political economy and culture when talking about epistemic virtues. Yet, this does not mean that we cannot address the subject. “I” can realize that “my” epistemic traditions are not the only traditions "I" should rely on, and epistemic modesty remains a value that "I" can choose, despite the realities of academia and culture. To do so will have more negative than positive consequences in the current academic landscape, and may require a subjectivity that embraces *courage*, a classical virtue (courage is not emphasized in academia either). On an analytical scale, however, the endorsement of virtues must be understood on the background of the dialectics of subjectivity and society.

## Conclusion

Focusing on one aspect of human subjectivity, namely epistemic virtues and values, demonstrates that the academic subject cannot be subtracted from the world. Yet, a theory of subjectivity also shows that academic subjects have agency inside and outside of their discipline, as narrow as this agency may be. Agency can take on different forms in different disciplines. I suggest that an analysis of epistemic modesty/grandiosity needs to be combined with the critical interests of the psychological humanities. This analysis may reach from when the subject of knowledge is demystified as a universal master-mind who is uninfluenced by extrinsic sources and immune to shortcuts in thinking and doing, to personal reflexivity and interference.

From a philosophical point of view, epistemic honesty requires the laying open of the sources of the limitations of knowledge, even when it is politically disadvantageous to do so, and even when it reinforces an attack on academia. Modesty does not entail the seeking only of a narrow-minded expertise beyond which one cannot contribute to the public debate. Yet, critical modesty demands that expertise is augmented with critical thinking, thinking that refuses to simply follow a neoliberal agenda, and that reflects on the assumptions, strengths, and weaknesses of science. Emphasizing reflexivity and interference as sources of strength for the sciences is something that critical modesty requires. Such critical reflection does not necessarily provide the assurance for a better science, but rather supports the conditions for its possibility.

History has always had a special status in the study of science, and it must also hold such status in psychology, when one traces the history of epistemic subjectivity. A psychology of science identifies how subjectivity has changed over time and how it shapes current research practices. Such a new psychology of science should not be developed in order to denounce science, but rather in order to identify its relevance for addressing current problems, as well as to reconceptualize problems and apparent solutions that have often hindered truthful action in the world. Psychologists of science, colleagues, and students of psychology should remain careful about researchers whose primary interest is producing, marketing, and selling a product, and who use neoliberal self-promotion to increase shares in the market place.

Beyond reflexivity, epistemic modesty requires interference not only in one's own epistemic shortcomings, which necessitates a constant improvement in terms of the processes and contents of knowledge, but also in terms of knowledge claims that others are making. If those claims are problematic from an epistemo-ethical perspective, then academic and public interrogations are required. The move from the ivory tower to recommendations for practice needs to involve reflection upon the concept of applicability, which needs to be challenged if it only means support for systems of power and financial interests. In this process, it is important that modesty does not become its opposite, the grandiosity of critique. Critique itself needs to remain modest if it seeks to do justice to its meaning, an equally difficult challenge in a neoliberal, post-truth context.

## References

Aalbers, D., & Teo, T. (2017). The American Psychological Association and the torture complex: A phenomenology of the banality and workings of bureaucracy. *Journal für Psychologie, 25*(1), 179–204.

Adams, G., & Estrada-Villalta, S. (2017). Theory from the South: A decolonial approach to the psychology of global inequality. *Current Opinion in Psychology, 18*, 37–42. https://doi.org/10.1016/j.copsyc.2017.07.031.

Bacon, F. (1965). *A selection of his works* (S. Warhaft, Ed.). Toronto, ON, Canada: Macmillan.
Bhatia, S. (2018). *Decolonizing psychology: Globalization, social justice, and Indian youth identities*. Oxford, UK: Oxford University Press.
Bourdieu, P. (1988). *Homo academicus* (P. Collier, Trans.). Stanford, CA: Stanford University Press.
Candolle, A. d. (1873). *Histoire des sciences et des savants depuis deux siècles: Suivie d'autres études sur des sujets scientifiques, en particulier sur la sélection dans l'espèce humaine*. Genève: Georg.
Chakrabarty, D. (2000). *Provincializing Europe: Postcolonial thought and historical difference*. Princeton, NJ: Princeton University Press.
Code, L. (1993). Taking subjectivity into account. In L. Alcoff & E. Potter (Eds.), *Feminist epistemologies* (pp. 15–48). New York: Routledge.
Daston, L., & Galison, P. (2007). *Objectivity*. New York, NY: Zone.
Devereux, G. (1967). *From anxiety to method in the behavioral sciences*. New York: Humanities Press.
Feist, G. J. (2006). *The psychology of science and the origins of the scientific mind*. New Haven, CT: Yale University Press.
Feyerabend, P. (1975). *Against method: Outline of an anarchistic theory of knowledge*. London: New Left Books.
Fine, G. (2008). Does Socrates claim to know that he knows nothing? *Oxford Studies in Ancient Philosophy, 35*, 49–85.
Fleck, L. (1979). *The genesis and development of a scientific fact*. Chicago, IL: University of Chicago Press (Original work published 1935).
Fowers, B. J. (2005). *Virtue and psychology: Pursuing excellence in ordinary practices*. Washington, DC: American Psychological Association.
Frankfurt, H. G. (2005). *On bullshit*. Princeton, NJ: Princeton University Press (Original work published 1986).
Freud, S. (1977). *Vorlesungen zur Einführung in die Psychoanalyse*. Frankfurt/Main: Fischer.
Galton, F. (1962). *Hereditary genius: An inquiry into its laws and consequences*. Cleveland, OH: World (Original work published 1869).
Graeber, D. (2011). *Debt: The first 5000 years*. London: Melville House.
Grenberg, J. (2005). *Kant and the ethics of humility: A story of dependence, corruption, and virtue*. New York: Cambridge University Press.
Guilford, J. P. (1950). Creativity. *American Psychologist, 5*(9), 444–454. https://doi.org/10.1037/h0063487.
Harding, S. (1986). *The science question in feminism*. Ithaca, NY: Cornell University Press.

Harvey, D. (2005). *A brief history of neoliberalism*. Oxford, UK: Oxford University Press.

Henrich, J., Heine, S. J., & Norenzayan, A. (2010). The weirdest people in the world? *Behavioral and Brain Sciences, 33*(2–3), 61–83. https://doi.org/10.1017/S0140525X0999152X.

Hersch, E. L. (2006). Philosophically-informed psychotherapy and the concept of transference. *Journal of Theoretical and Philosophical Psychology, 26*(1–2), 221–234. https://doi.org/10.1037/h0091276.

Hiebsch, H. (1977). *Wissenschaftspsychologie: Psychologische Fragen der Wissenschaftsorganisation*. Berlin: Deutscher Verlag der Wissenschaften.

Holton, G. (1973). *Thematic origins of scientific thought: Kepler to Einstein*. Cambridge, MA: Harvard University Press.

Horkheimer, M., & Adorno, T. W. (1982). *Dialectic of enlightenment*. New York, NY: Continuum (Original work published 1947).

Jackson, J. P., & Weidman, N. M. (2004). *Race, racism, and science: Social impact and interaction*. Santa Barbara, CA: Abc-Clio.

Keller, E. F. (1985). *Reflections on gender and science*. New Haven, CT: Yale University Press.

Kim, U., Yang, K.-S., & Hwang, K.-K. (2006). *Indigenous and cultural psychology: Understanding people in context*. New York: Springer.

Kuhn, T. S. (1962). *The structure of scientific revolutions*. Chicago: University of Chicago Press.

Lombroso, C. (1905). *The man of genius* (2d ed.). London: W. Scott.

Marx, K. (1962). *Das Kapital: Kritik der politischen Ökonomie (Erster Band) (Marx Engels Werke Band 23)* [Capital: Critique of political economy (Volume I) (Marx Engels Works: Volume 23)]. Berlin: Dietz (Original work published 1867).

McIntyre, L. (2018). *Post-truth*. Cambridge, MA: MIT Press.

Müller-Freienfels, R. (1936). *Psychologie der Wissenschaft*. Leipzig: Barth.

Oeser, E. (1988). *Das Abenteuer der kollektiven Vernunft. Evolution und Involution der Wissenschaft*. Berlin: Parey.

Oreskes, N., & Conway, E. M. (2010). *Merchants of doubt: How a handful of scientists obscured the truth on issues from tobacco smoke to global warming*. New York: Bloomsbury Press.

Osbeck, L. M., Nersessian, N. J., Malone, K. R., & Newstetter, W. C. (2011). *Science as psychology: Sense-making and identity in science practice*. New York: Cambridge University Press.

Ostwald, W. (1908). *Erfinder und Entdecker*. Frankfurt am Main: Rütten & Loening.

Plato. (1997). *Complete works* (edited, with introduction and notes by J. M. Cooper; associate editor, D. S. Hutchinson). Indianapolis, IN: Hackett.

Popper, K. R. (1972). *Objective knowledge: An evolutionary approach*. Oxford: Clarendon Press.

Popper, K. R. (1992). *The logic of scientific discovery.* London: Routledge (Original work published in 1935).

Proctor, R. N., & Schiebinger, L. (Eds.). (2008). *Agnotology: The making and unmaking of ignorance*. Stanford, CA: Stanford University Press.

Reichenbach, H. (1938). *Experience and prediction: An analysis of the foundations and the structure of knowledge*. Chicago, IL: The University of Chicago Press.

Schraube, E. (2013). First-person perspective and sociomaterial decentering: Studying technology from the standpoint of the subject. *Subjectivity, 6*(1), 12–32. https://doi.org/10.1057/sub.2012.28.

Simonton, D. K. (1988). *Scientific genius: A psychology of science*. New York: Cambridge University Press.

Smyth, J. (2017). *The toxic university: Zombie leadership, academic rock stars and neoliberal ideology*. London, UK: Palgrave Macmillan.

Spivak, G. C. (1999). *A critique of postcolonial reason: Toward a history of the vanishing present*. Cambridge, MA: Harvard University Press.

Teo, T. (2008). From speculation to epistemological violence in psychology: A critical-hermeneutic reconstruction. *Theory & Psychology, 18*(1), 47–67. https://doi.org/10.1177/0959354307086922.

Teo, T. (2016). Embodying the conduct of everyday life: From subjective reasons to privilege. In E. Schraube & C. Hojholt (Eds.), *Psychology and the conduct of everyday life* (pp. 111–123). London: Routledge.

Teo, T. (2017). From psychological science to the psychological humanities: Building a general theory of subjectivity. *Review of General Psychology, 21*(4), 281–291. https://doi.org/10.1037/gpr0000132.

Teo, T. (2018a). *Outline of theoretical psychology: Critical investigations*. London, UK: Palgrave Macmillan.

Teo, T. (2018b). Homo neoliberalus: From personality to forms of subjectivity. *Theory & Psychology, 28*(5), 581–599. https://doi.org/10.1177/0959354318794899.

Tucker, W. H. (1994). *The science and politics of racial research*. Urbana, IL: University of Illinois Press.

Wertheimer, M. (1945). *Productive thinking*. New York: Harper.

Wilkinson, R. G., & Pickett, K. (2009). *The spirit level: Why more equal societies almost always do better*. London: Allen Lane.

# 3

# Rethinking Psychology of Technology for Future Society: Exploring Subjectivity from Within More-Than-Human Everyday Life

Niklas Alexander Chimirri and Ernst Schraube

Two decades into the twenty-first century, the gap between techno-scientific progress and an understanding of its significance in human life seems wider than ever. Günther Anders, a philosopher of technology, wrote over 60 years ago about an increasing discrepancy, a "Promethean gap" (Anders, 2018a/1956, p. 29), between human creation and imagination. Through the development of modern technology, he realized, human activity had begun to surpass itself in a problematic way. Since human capacities such as emotion, perception, or even the ability to care are relatively limited when compared to our capacity of making, we are faced with a fundamental discrepancy between the world of technology and the human ability to meaningfully conceive it; a divide

N. A. Chimirri · E. Schraube (✉)
Department of People and Technology, Roskilde University, Roskilde, Denmark
e-mail: schraube@ruc.dk

N. A. Chimirri
e-mail: chimirri@ruc.dk



K. C. O'Doherty et al. (eds.), *Psychological Studies of Science and Technology*, Palgrave Studies in the Theory and History of Psychology,
https://doi.org/10.1007/978-3-030-25308-0_3

primarily attributable both to the accelerated pace of technological development, and to the enormous complexity of the things created and their effects. In this paradoxical situation, whereby "we are smaller than ourselves" (Anders, 2018b, p. 324, authors' translation), Anders sees the basic contradiction of our time and the decisive task to situate ourselves, our ways of thinking, our theories, interpretations and actions, within the horizon of the self-created world of high technology. "If we don't succeed", he underlines, "in matching the circumference of our imaginative abilities with our abilities of making, then we won't survive" (1992, p. 8, authors' translation).

Today, scientists around the globe are increasingly aware how the world is dangling on a string due to excessive human exploitation of the Earth's ecosystems, and are warning that the "time is running out … soon it will be too late to shift course away from our failing trajectory" (Ripple et al., 2017, p. 1028, in the declaration *Warning to Humanity: A Second Notice*, signed by over 15,000 scientists). Within Science and Technology Studies (STS) influential scholars realize this danger as well, and put it at the center of their thinking (e.g., Haraway, 2016; Latour, 2017; Papadopoulos, 2018; Stengers, 2015). As a recent response, a critical self-reflection has set in among scientists in various disciplines on how modern sciences have been part of the problem, as well as on how we have to fundamentally rethink our scientific conceptions and self-understandings—to be able to meaningfully work with these problems and thus become part of enabling a viable future society. With reference to the body of work of Günther Anders, Bruno Latour, for instance, explicates to his fellow scholars in the light of the threat of global warming: "You are interesting to me only if you situate yourselves *during* the end time, for then you know that you will not escape from the time that is passing. Remaining in the end time: this is all that matters" (2017, p. 187).

Likewise, psychologists are concerned and engage in a critical debate about their scientific apparatus and a renewal of psychological theory, methodology, and research practice. In this paper we expand this dialogue with STS to rethink the psychological study of technology. Our question is how the psychology of technology can fundamentally reformulate its scientific vision so it can help to analyze the discrepancy between creation and imagination, and thereby contribute to a profound understanding of the significance of modern technology in human life. We are arguing for a conceptual shift along four lines: (1) From a disembodied, dissecting, and individualizing scientific vision of subjectivity toward an embodied conception of the internal relationship between humans and the more-than-human world; (2) from an external and artificially distancing "view from above", including a subduing research practice, toward restructuring research from a situated standpoint of the human subject; (3) from "methodolatry" (Bakan, 1967) and its quick-fix methodical recipes toward content-based methodologies enabling the exploration of the complexity and conflictuality of the internal relationship between humans and the world; and finally (4) from conceptualizing technology as neutral instruments for controlling world toward grasping technological artifacts as contradictory and political forms of everyday life.

Along and across these lines we unfold in the following the contours of a conceptual renewal and the perspective of a critical participatory psychology of technology from the standpoint of the human subject. Subjectivity, so we argue, needs to be decentered, by understanding it as not exclusively belonging to individual human beings. Rather, subjectivity is done from within its more-than-human relations. Here we build on posthumanist thought and STS scholars, who situate human action in the vision of "more-than-human worlds". We build on this term, because, as Maria Puig de la Bellacasa explains, "it speaks in one breath of nonhumans and other than humans such as things, objects, other animals, living beings, organisms, physical forces, spiritual entities, and humans" (2017, p. 1) and renders it possible to question

> the boundaries that pretend to define the human realm (against the other than human as well as otherized humans), to sanction humanity's separate and exceptional character and, purposely or not, to sanction the subjection of *everything else* to this purported superiority. The frontiers blurred through these ways of thinking and the sociomaterial moves that impel them are now commonly known: between nature and culture, society and science, technology and organism, humans and other living forms. (Puig de la Bellacasa, 2017, p. 12)

At the same time, this decentered relational human subject always already acts from a particular standpoint, from within its experience of everyday life. Accordingly, all knowledge, including all psychological knowledge, emerges from acting from within this more-than-human everyday life.

Psychological concepts must be able to grasp subjectivity in this contradictory movement, as both being decentered and dependent on more-than-human worlds, and simultaneously as concretely situated within the embodied experiential realm of human everyday life. As will be shown, this is a crucial step toward overcoming psychology's disembodied, dissecting, and atomizing conception of the human–technology relation.

## Building a Psychology of Technology for a Sustainable Human–World Relationship

The psychological study of technology can play its part in investigating the discrepancy between creation and imagination by rethinking a scientific culture that situates the study of the relationship between humans and technology within more-than-human everyday life experienced and lived. Such a psychology regards the crucial challenges of today's technological world, for instance social inequality or climate change, as inextricably entangled with how we conduct our everyday life, and with how we come to know about and understand this ontological, epistemological, and ethical entanglement. It questions a top–down, instrumentalist scientific gaze from outside and above, by systematically including the researcher's internal relatedness to the subject matter, given that her knowledge creation and subjectivity are just as much rooted in the researcher's everyday entanglement with the technological world as any other's. Psychological knowledge thus becomes dependent on the development of self-understanding, which is generalized via a collectivized, critical exchange of everyday experiences with technology across the shifting standpoints of people involved.

On the grounds of such a critical, self-critical, and participatory inquiry, psychology could be rendered able to meaningfully engage in current debates on how scientific knowledge can contribute to maintaining human life by building a human–world relationship that is worthwhile sustaining for every organism inhabiting planet Earth. It could thus contribute to recent debates that seek to fundamentally reconceptualize humanity's inextricable relatedness to the world, debates that for instance draw on and develop concepts such as the Biosphere, Gaia, the Anthropocene, Terrapolis, the Chthulucene, etc. Such concepts invite explorations across STS, the natural or "zoe" sciences, as well as the (post-) humanities more broadly (as, for example, discussed in the *South Atlantic Quarterly* special issue 2017 on Climate Change and the Production of Knowledge, edited by literary scientists Ian Baucom and Matthew Omelsky). What role can psychology play in this transdisciplinary project of fabricating an alternative view of the human–world relationship, and in that context, of our relationship to technology?

Baucom and Omelsky describe science's current transdisciplinary challenge as follows:

> Is it possible to imagine a reinvention of feedforward possibilities, a reimagination and a new fabrication of some point of feedforward vitality from the conjoined perspectives of the human, social, and natural sciences? Can we fashion a perspective on the Anthropocene that is somehow both within it and at some (seeming) critical distance from it, a perspective through which we *can* "mobilize" our knowledge of having come to this point in the history of knowledge and, so, also mobilize the form that knowledge and the imagination can now take for the future of the planet? (2017, p. 15)

Psychology can be pivotal for addressing the consequences of our own creations to help us grasp the ambivalences and contradictions we see ourselves confronted with in today's scientific and technological world, and to concretize and develop feedforward vital possibilities for reimagining, fabricating, and mobilizing situated and yet self-critical knowledge of daily existence. However, this requires acknowledging that the human–science–technology relation is contradictory because we are not mere victims of our past technological decision-taking, of our former designs and current use. We are fabricators of our daily technological reality, and we embody technological artifacts because we hope they will improve our life circumstances, that artifacts will help us expand our agency by rendering us able to grasp the world better—arguably at "a (seeming) critical distance", to paraphrase Baucom and Omelsky. Technology is never only detrimental, never only worsening our life circumstances: technology has also been co-constitutive for human development, for developing from primarily reactive organisms to active provision-producing, societally arranging beings—in order to create an alienating and simultaneously emancipating distance from vital daily necessities (Marvakis, 2013). Technology thus holds potentials for overcoming an immediacy-fixated reactivity to an individual's environment, by materially generalizing social possibilities for more purposefully acting in the world together; and it co-constitutes the core of human social self-understanding and the ways in which we think and act in everyday life.

Psychology's hitherto understanding of technology, in contrast, has been reproducing the common idea that technology is merely an external tool at the mercy of individuals' and cultures' will and hands. Its deeply contradictory and ambivalent nature has been ignored, in particular to what extent it is taken-for-granted, embodied, fully entangled in how we imagine and practice our daily existence—reciprocating all those societal contradictions the human-world relationship has historically come to produce. Currently, psychology's methodolatry de facto spurs an immediacy-fixated, instrumentalist, and individualistic understanding of science and technology, which promotes anthropocentrism and human (and Western) exceptionalism, and regards the world including human life as a mere resource for technological extraction and industrial exploitation through hegemonic political and economic forces (Zuboff, 2019). Instead of looking at the internal relatedness of the human–technology relationship, it detaches its study of the human and technology from everyday life lived, from the concrete everyday life contexts from within which it takes place. It atomizes its respective insights rather than considering them as entangled with and dependent of other aspects of daily human existence. It is this uncritical, detaching, anthropocentric understanding of science and technology that will not be able to tackle the Promethean gap that Günther Anders identified (on the contrary, it exactly creates it), and that we therefore suggest to replace with a psychology of technology situated in the standpoint of the human being and in its more-than-human practice of everyday living.

## The Need for Overcoming Psychology's Disembodied Conception of the Human–Technology Relation

In the 1970s, a sensitivity emerged within psychology toward the more questionable and contradictory aspects of techno-scientific progress. Various voices within the field began to call for a *psychology of technology*. Actually, Günther Anders was one of the first to argue for a systematic psychology of technology. In a paper written in 1961, he reflects on the implications of technology in people's everyday life, articulating the "need for a special psychological discipline" focusing on material objects (2018b, p. 60, authors' translation). Referring to the work of his father, William Stern (one of the most influential German psychologists in the first part of the twentieth century), Anders notes:

> My father has coined the unfortunate term "psychotechnics", although he didn't boast, as did his colleagues, to have discovered that the mind can be treated technically. In contrast, if we talk about "psychology of technology" we mean the study and critique of the existing influence that technology has on humans. (Anders, 2018b, p. 464, authors' translation)

Moreover, arguments for a psychology of technology were also put forward from within psychology. Kenneth Gergen, for example, explains: "We rapidly assimilate new technologies into our lives; we welcome and embrace them. But too seldom do we ask questions about the ways they have changed our lives – sometimes irrevocably" (2000, p. xiii). Various voices explicitly argue for the need of a psychological study of technology, since there would be hardly any area within psychology in which technology is not involved; be it the psychology of development, psychology of personality, social psychology, educational psychology, or work and organizational psychology (e.g., Gordo-López & Parker, 1999; Turkle, 2015; Walkerdine, 1997). Psychologists however also realized how the study of technology would be a real challenge for the whole discipline. Regina Becker-Schmidt, for example, writes: "The influence of the technological revolution on the bodily, psychosocial and mental constitution of whole generations has been ignored"

(1989, p. 50), and she emphasizes that we must be prepared to "explore an unknown scientific continent" (p. 49, authors' translation).

Despite the challenges psychology faces in developing a psychology of technology, the necessity of such a program is now widely acknowledged. Furthermore, in recent years various approaches have been evolving more on the margins of dominant psychology—ranging from psychoanalysis, the cultural-historical activity theory, to cultural psychology, discursive psychology, social constructionism and critical psychology—to a quite substantial body of thought examining in-depth the significance of technological artifacts in human experience and action. Yet, in relation to the amplitude of academic psychology and its dominance in everyday societal discourse, these approaches remain the exception.

Reasons for why the psychology of technology has, as of yet, so tentatively developed in psychology can be found in modern psychology's proximity to discourses promoting techno-scientific progress, and, above all, in its one-sided conception of itself as a science mainly rooted in quantitative methodology. Since psychology detached itself from philosophy in the mid-nineteenth century and was institutionalized as an independent scientific discipline, major traditions of psychological thought adopted their theoretical language and methodology from the natural and computer sciences (using terms such as "input", "output", "storage", etc.). The methodological core of its scientific identity lies in the acquisition of knowledge through quantitative, experimental, and statistical procedures.

Such an approach brought forth two fundamental problems. The first is a problem situated in the realm of scientific theory. Rather than developing methodology according to the investigated subject matter—adequate to the psychological phenomena under scrutiny—the quantitative, experimental, and statistical methodology is assumed as valid in advance, independently of the content and context of research. Moreover, this is viewed as a universal method of acquiring knowledge—a methodological fallacy which by no means guarantees the desired scientific objectivity, and which has repeatedly been critically addressed, for example, as "methodolatry" (Bakan, 1967) or "methodologism" (Teo, 2009). This brings with it, second, a substantive problem related to the subject matter of research. On the foundation of quantitative methods and a mechanistic scientific language adopted from natural and engineering sciences, psychological phenomena become reduced to simplistic cause and effect models. In such an approach, human experience, subjectivity, life contexts, and potential ways of realizing human agency can only be understood in abstract ways: as artificially dissociated variables that operationalize and thereby unduly reduce the complexity of a person's whole relations to the world. The internal relatedness and two-sidedness of the human–technology relationship becomes disembodied and dissected, while its phenomenal expressions are atomized by removing them from the everyday life and contexts in which they actually unfold, are lived and practiced.

The problem of methodologism and an ensuing lack of ecological validity has also been identified *within* classical psychology's few debates on the relevance of psychology of technology. For example, Walther Bungard and Jürgen Schultz-Gambard explain self-critically that in dealing with technology, psychology's quantitative methodology raises its own barriers. As they emphasize, an epistemology which reduces the reality under investigation to simple cause and effect models creates a "decontextualized psychology", where technology can hardly appear as a promising research object (Bungard & Schultz-Gambard, 1988, p. 161). This illustrates how the question of the psychological study of technology exposes the limits of a psychology that theoretically and methodologically reduces the notion of the psyche to mechanical and technical terms, and underscores the need for working on a fundamental renewal of psychology's understanding of itself as a science.

In this sense, not only is the critical analysis of technology and scientific technicism integral to the production of psychological knowledge and thereby developing the field of psychology of technology. In addition, efforts have to be made to develop new theoretical concepts and methodologies for conducting empirical research. These concepts are to transgress detached, isolating, and individualizing ways of understanding the psyche as a mechanical process, and which instead regard each person as a subject actively experiencing and acting *from within* a world co-constituted and mediated by technology (Schraube, 2019).

In this more dialectical line of theorizing the human–technology relationship, more recent psychological theories emphasize the efficacy of objects and the "materialized action" of technological artifacts (Schraube, 2009). This aspect is highlighted, for instance, in the notion of objects as "affordances" (Gibson, 1986), or the idea of conceptualizing technological things as "evocative objects" (Turkle, 1984), as "actors" (Latour, 2005) or as "political forms of life" (Feenberg, 2017; Winner, 1989). Even if the activities embodied in the things always contain an objectivized and generalized dimension—with Elaine Scarry, for instance, talking of manufactured artifacts as "compassion-bearing" objects (1985, p. 293)—the objectification movement, due to its specific, situated societal and historical character, cannot entirely avoid a one-sided, fractured, and partial character. Furthermore, an object can never be designed in a way that it can only be used for its originally intended purpose. Therefore, technological objects represent fundamentally contradictory and conflictual things. This inherent contradictoriness and relational-interpretive openness of technology—and herewith of psychological science, which necessarily relies on technology in the form of scientific objects and methods—must be integrated in psychology's understanding of science and technology as inextricably entangled part of human imagining and acting in more-than-human worlds. This encompasses the researcher's imagining and acting just as much as that of anyone else.

## The Internal Relatedness of Science, Technology, and the Researcher

Sharing the commitment of developing theoretical and methodological frameworks that attempt to investigate human–technology relations from within the more-than-human world, a variety of psychological approaches have emerged (as mentioned above) that engage in unfolding a *psychology of technology*. Even if their points of discursive interaction still tend to be rather sporadic and unconnected, all are working on a range of common concerns: from a fundamental renewal of psychology's understanding of itself as a science, toward situated, qualitative approaches driving psychology's production of knowledge and research practice in the exploration of the human–technology relationship from within the everyday life it takes place in.

The various approaches to psychology of technology build, on the one hand, on the wealth of diverse traditions of qualitative research within the social sciences and humanities, developing them in accordance with specific research topics and questions (at times including quantitative methods for particular issues). On the other hand, the psychological study of technology is closely related to and involved in the development of STS. STS is an inter- or rather transdisciplinary field of study, bringing together various traditions of thought including philosophy, sociology, history, anthropology, political science, and psychology. Over the past decades, STS has established itself at European and North-American universities as an independent field of research and teaching, systematically investigating and debating the relationship between science, technology and society.

On the basis of detailed analyses of the production of scientific knowledge (Knorr-Cetina, 1981; Latour & Woolgar, 1986), STS argues to move beyond a positivist, cause-effect-model seeking conception of scientific research practice. In addition, it calls for a notion of science and technology as socially situated processes that can only be adequately grasped through their material, cultural, and social entangledness. Accordingly, the empirical research into the relations between science, technology, human agency and life, builds especially on qualitative methodologies, which investigate human language and action as they are practiced in everyday life (Hasse, 2015; Hess, 2000; Schatzki, Knorr Cetina, & von Savigny, 2001). A constituting element of the formation of STS consists in an increasing awareness of a profound crisis in modern life and thought. As Langdon Winner explains in an analysis of basic concerns and projects that have inspired research and thinking in STS during the past several decades:

> A fourth collection of concerns in STS attracts philosophers and social theorists. Here the focus turns to what many thinkers have argued is a profound crisis in the underlying conditions of modern life and thought. The development of modernity has gone badly wrong, not only at the

> level of specific, vexing social problems but in its fundamental core of ideas and institutions, especially those that involve science and technology. While attempts to fathom the nature of the crisis vary from writer to writer – from Marx to Mumford, from Heidegger to Ellul, from Habermas to Foucault – the point of inquiry is to locate philosophical, historical, and cultural origins of phenomena closer to hand. In its very nature, research of this kind is both radical and critical; it seeks to "look deeper", to probe what may be highly general sources of contemporary disorientation and to suggest change of the most fundamental kind. (Winner, 1996, p. 104)

Major traditions in STS realize that modernist understandings of science and technology need to be transgressed. Scientific modernism propels anthropocentrism and human exceptionalism, which puts human beings first in a presumed natural hierarchy, and the world and its resources at the extractable service of humanity. Implied is an artificial detachment of humanity from this very world that human beings are also a part of, which they are intra-dependently related with, or which, in Annemarie Mol's terminology, *transubstantiates* them (Mol, 2008).

While, for instance, the cultural-historical school of psychology represents a practice-based, dialectical approach that engages in the study of technology (we will discuss it in more detail below), its ontological framework seldom explicitly questions human exceptionality and supremacy. Here, technological things still tend to be conceptualized as either "tools" or "resources", both of which connote instrumentalist understandings of material objects—and may thereby overlook the complexity, contradictoriness, and politics of technological artifacts.

Dialectical approaches to the study of science and technology thus just as much require a rethinking of their ontological framework, for instance toward a dialectical psychology based on a *philosophy of internal relations* (Ollman, 2003, 2015). Ollman's proposition to study "contradictions" as an alternative to isolating essences and dichotomizing phenomena that are internally related, addresses concerns similar to those articulated by more posthumanist approaches, such as by the methodology of *diffraction*, which quantum physicist and feminist theorist Karen Barad developed further on the basis of Donna Haraway's work (Barad, 2007, 2014). In the words of Barad:

> Diffraction is not only a lively affair, but one that troubles dichotomies, including some of the most sedimented and stabilized/stabilizing binaries, such as organic/inorganic and animate/inanimate. Indeed, the quantum understanding of diffraction troubles the very notion of dichotomy – cutting into two – as a singular act of absolute differentiation, fracturing this from that, now from then. (Barad, 2014, p. 168)

According to our reading of Ollman's dialectics through Barad's diffractive methodology, scientific analysis and research practice is never a process which a subject individually engages in: she is always already internally related to (or entangled with) the more-than-human world, as part of an apparatus that consists of a multiplicity of forces that cogenerate the "result" of an analysis. Analysis always cuts together-apart in different ways, is inherently contradictory because it makes some things emerge while simultaneously shunning others—and it is in the difference of these analyses that the potential for purposeful collective action emerges. That which is temporarily shunned from the analysis is also part of its mutually dependent relations: what can be seen and researched, what is present in our analyses, can only be foregrounded because all else is *absenced* (Law, 2007). And yet, also the absenced background is present, and co-constitutes what can be researched. It is part of the whole subject matter under scrutiny, for instance of "subjectivity". Thus the whole is always already radically situated: only very particular and partial knowledge about it can be generated. Knowledge's radical situatedness must be rendered as explicit as possible, opening itself up for a critical inquiry by other apparatuses that analytically cut together-apart differently, presence and absence other aspects of the internal relationship of the whole phenomenon under scrutiny.

In order to grapple with the particularity and partiality of a researcher's analysis as situated in an entangled apparatus, the question of the *self-reflexivity of research in the psychology of technology* has become one key element in current debate, as the researcher also needs to situate her own inquiry in the internal relatedness and contradictoriness of human–technology relations. Svend Brinkmann underlines the need to take the everyday life of researchers, including their things and situatedness, as the starting point of qualitative research: "Everyday life *objects* are thus those that the researcher in question appropriates and uses in her daily living (e.g., consumer products, technologies, pieces of art), and everyday *situations* and *events* are those that the researcher experiences in her life (e.g., conversations, parties, work, rituals)" (Brinkmann, 2012, p. 17). The central relevance of analyzing this self-involvement, something which various feminist theorists have explicitly put on the research agenda in the past decades, is also evident in the psychology of technology. As Barad has noted, ontologically we are not merely in the world, but *of* the world (Barad, 2007). Hence, we are also permeated by precisely that technology we have created and create, and which influences the everyday lives of all of us (Højgaard & Søndergaard, 2011; Ingold, 2013). Epistemologically, this implies that we cannot locate ourselves as researchers as detached from or instrumentally outside our subject matter, but instead inside the situatedness of human relations to technology. As Donna Haraway emphasizes, the view from outside, the "infinitive vision" of human relations to technology, is a fiction: "Only partial perspective promises objective vision" (1991, p. 190).

Meanwhile, it is precisely the individualizing optical metaphors of *reflection* and *self-reflexivity*, so ubiquitously used in the qualitatively working social sciences and humanities as criterion for ensuring validity and objectivity, that Haraway (1997) later questioned by offering the concept of *diffraction*. As Vivienne Bozalek and Michalinos Zembylas explain:

> Reflection remains fundamentally an inner mental activity in which the researcher supposedly takes a step back and reflects at a distance from the outside of the data … Reflection is thus based on the assumption of an 'I' who is different and exterior to that which is conceptualizing, an 'I' who is separate from the world … The slip into the subject 'I' is important in understanding reflection and diffraction, since in the latter there is no researcher as independent subject – in diffraction the intra-action and connections between human and non-human phenomena are foregrounded. Rather than pondering on the meaning of texts or events, a diffractive methodology focuses on what these phenomena do and what they are connected to. (2017, pp. 116–117)

In order to focus on what phenomena do and what they are connected to, the method of diffraction, as science philosopher Melanie

Seghal points out, "incorporates historicity and difference into the practice of theory itself" (2014, p. 188). This systematic incorporation of historicity and difference in the practical-performative act of analyzing and thus producing knowledge across human and more-than-human connections echoes, as we will in the last section argue, also a central tenet of dialectical, practice-based approaches to a psychology of technology. It aims at the transgression of particular and only partial perspectives on internal human-technology relatedness through the generalizing creation of knowledge, while its generalizations highlight connections alongside their contradictoriness and difference, thus remaining open to ongoing renegotiation.

As Seghal (2014) further points out, Barad develops Haraway's notion of diffraction by emphasizing its ontological implications and its internal relatedness (or entangledness) with epistemology, methodology, and ethics. Similarly, we will argue that a *diffracting dialectics* can only be developed on the ground of profound ontological reformulations and refinements, most importantly of the researcher-researched relationship—as the necessarily particular and partial, limited analysis of the internal relatedness of human–technology–world only becomes truly open to difference, becomes questionable and negotiable via everyday practice, if troubled in its most fundamental assumptions.

Simultaneously, the troubling also requires a counter-movement, an at least temporary agreement across diverse and necessarily limited perspectives and actions, in order to render collaboration and thus coexistence (across human and nonhuman, or more-than-human forces) possible. Temporary agreement in the form of conceptual generalizations that do not deadlock human–nonhuman–technology–world relations and intralink empirical findings, we argue, is what the psychology of technology from the standpoint of the subject particularly works toward. Subjectivity, as the most central of all concepts in this psychological tradition (Teo, 2017), is here understood as the conduct of everyday life, which necessarily bridges across and integrates the various practices, contexts, viewpoints, collectives that a human being contributes to. Collectively troubling this integration by creating difference, by critically inquiring into the taken-for-granted, is a vital part of everyday life. But also this troubling requires generalization, in order to be able to "stay with the trouble", in paraphrasing Haraway (2016), to acknowledge but not merely reproduce Anders' Promethean gap, but to diffract it. It requires explicitly working toward expanding human experiencing and self-understanding, in terms of its internal relatedness with the technologized world, thereby rendering purposeful collaboration on the societal and ecological crises of our time with more-than-human forces increasingly possible.

## Toward a Critical-Participatory Psychology of Technology from the Standpoint of the Human Subject

As part of the earlier mentioned special issue on *climate change and the production of knowledge*, feminist philosopher Rosi Braidotti (2017) writes about how knowledge creation must be understood as a deeply political endeavor, which challenges the fracturing of human–nonhuman relations via negative and deadlocking differentiation—a fracturing that, as we have argued, has been strongly propelled by modern psychology with its many disembodying, dissecting and atomizing conceptualizations of subjectivity that create an artificial distance to subjectivity's more-than-human world relations, including its technological relations. In its stead, a psychology that builds on the feminist and posthumanist critiques raised in the past few decades and that aims at overcoming instrumentalist-exploitative understandings of the world at the technological service of supreme human beings, must radically situate itself and its inquiries in the everyday practice of diverse and critical knowledge creation—a knowledge created together including all those missing humans that else are otherized, overlooked, differentiated away, in the collaborative actualization of possible (and vitally more sustainable) futures. Psychologists, along other scientists, need to acknowledge the partiality and particularity of their theorizing, thus becoming critical subjects of knowledge. Our claim resonates with Braidotti's perspective when she emphasizes:

> The task of critical subjects of knowledge is to pursue the posthuman, all-too-human praxis of speaking truth to power and working toward the composition of planes of immanence for missing peoples, respecting the complex singularities that constitute our respective locations. "We" is the product of a praxis, not a given. The dwellers of this planet at this point in time are interconnected but also internally fractured by the classical axes of negative differentiation: class, race, gender and sexual orientations, and age and able-bodiedness continue to index access to normal humanity. This rhizomic field of posthuman knowledges does not aspire to a consensus about a new humanity but labors to produce a workable frame for the actualization of the many missing people, whose "minor" or nomadic knowledge is the breeding ground for possible futures. (Braidotti, 2017, p. 93)

The critical-political knowledge practice that Braidotti argues for requires, in our understanding, what Baucom and Omelsky called "a perspective on the Anthropocene that is somehow both within it and at some (seeming) critical distance from it" (2017, p. 15)—a distancing that is however neither artificial nor abstract, but concrete in that it serves affirmative collaboration for "vital geocentrism" (Braidotti, 2017, p. 91) rather than for conceptually detrimental anthropocentrism. We consider such a move to be a deeply dialectical (and arguably dialectic-diffractive) endeavor: conceptual practice should make analytical distancing possible, but not for abstract and solipsist, artificially detached views of subjectivity, but rather for situated concretization of partial and particular knowledge claims that seek more general directionalities for fellow action without fixating them. Generalizing concepts serve to open up for difference, for negotiating what to move toward how—to get at a distance from one's own doings in order to open up for questions of others' doings (Langemeyer, 2019, in this volume). It is this immanent contradictoriness of relational knowledge, as both seeking particularity and generalization (Dreier, 2007), or diffraction and affirmation (Haraway, 2016; Thiele, 2014), that we consider to be deeply dialectical, and that a conceptualization of subjectivity must render possible to emerge in its concrete-empirical actualizations.

But how to strive for the dialectical contradictoriness and open-endedness, the posthuman uncertainty and indeterminacy of the analytical work of each particularly and partially analyzing apparatus? Can it only be acknowledged, or can we actively work with that as a productive point of departure for developing a vitally *geocentric* psychology of science and technology? The solution may be to radically situate psychological inquiry in concretely experienced, contradictory everyday life, and to collectively develop conceptual as well as technological artifacts that incorporate a generalized human standpoint from within more-than-human everyday life: a cross-apparatus subjectivity that truly can serve an intra-dependent, vitally geocentric practice. Ergo: concepts that enable a practice of mutual recognition in the processes of making and transforming the world, by challenging and developing everyday self-understandings in dialogue with more-than-human forces.

This is what dialectically grounded, action-oriented understandings underline, which regard the human being as actively constituting and simultaneously constituted by her more-than-human environment and therewith technology. As mentioned above, however, dialectical psychological traditions are not free from reproducing modernist-instrumentalist conceptualizations of technology and world, without explicitly questioning human exceptionalism and supremacy. Drawing on the discussion in STS and in particular feminist notion of *diffraction* can help to specify how a dialectical approach to human-technology inquiry could more clearly address the internal relatedness and entangledness of human and world. In consequence, dialectical practice approaches to a psychology of technology require a (self-)critical and inherently participatory stance due to their acknowledgment of one's analyses' unavoidable particularity and partiality. But what role can "subjectivity" as an integrating and yet troubling concept play in overcoming instrumentalist understandings of science and technology? A look into the history of dialectical theorizing in psychology that has been developing concepts for investigating technological practice will bring us here a step further on.

Emerging in the early twentieth century and still influential today, the *cultural-historical tradition of psychology* has systematically studied the significance of material objects in human subjectivity. Here, the focus is especially on the development of human activity (Stetsenko,

2005). This tradition of thought is based on the assumption that higher psychological processes emerged in phylogenesis simultaneously with the capacity to produce and modify material objects as a means of regulating human interactions with the world and one another, and that this development was a prerequisite for the formation of human personality. As Alexander Luria wrote in 1928, "Man differs from animal in that he can make and use tools ... [Such tools] not only radically change his conditions of existence, they even react on him in that they effect a change in him and his psychic condition" (Luria, 1928, p. 493). Through the concept of objectification, the human production process is understood as a societal process of human externalization in the goods produced, into which flows the dimension of psychological content, such as human experience, needs and knowledge gained through involvement and conflicts with the natural and social world (Leontyev, 1981). For this reason, the phenomena are studied in situated ways both socially and historically, and issues of the democratization and the contradictory generalization processes inherent to objectivation processes are also a key theme in cultural-historical theory (Wartofsky, 1979). In particular, though, the central question in this approach is how the produced things are integrated as tools and means into goal-directed human action. On the grounds of the concept of objectification, the objects produced are understood as having both a material as well as an ideal dimension. Michael Cole has described this dual character of things as follows: "By virtue of the changes wrought in the process of their creation and use, artifacts are simultaneously *ideal* (conceptual) and *material*. They are ideal in that their material form has been shaped by their participation in the interaction of which they were previously a part and which they mediate in the present" (Cole, 1996, p. 117). In empirical research, one finds detailed models of how material objects co-constitute human activities, whereby the focus of such research is primarily on human development, learning and educational practices, as well work, design and organizational practice (Bang, 2012; Engeström, Miettinen, & Punamäki, 1999; Kaptelinin & Nardi, 2006; Kontopodis, Wulf, & Fichtner, 2011).

*Critical psychology* builds on ideas promoted in the cultural-historical approach in order to propose a psychology from the standpoint of the subject, which also formulates a theoretical and methodological foundation of a dialectical psychology of technology. One of the most important achievements of critical psychology consists in developing a psychological vocabulary articulating in detail the internal relationship between humans and the world. Human beings are not regarded as abstract, isolated individuals, but understood as unfolding their everyday life in relation to nature, culture, technology, and society—an entanglement where the concepts of human subjectivity, agency and the conduct of everyday life are pivotal.

In his historical analysis of the psyche, Klaus Holzkamp takes Leontyev's work as a starting point to elaborate an understanding of the crucial role of the human capacity to produce things, as well as the utilization of the tools and means produced, in creating the potential for human social existence (Holzkamp, 1983). Moreover, in extensive analyses, he highlights the problem of an instrumentalist scientific language in psychology and engaged in a fundamental renewal of the epistemological foundation of psychology (Holzkamp, 1983, 2013a, 2013b). A key moment in this renewal came with the realization that instead of human subjectivity and agency being causally *determined* by social and technological conditions, they are *grounded* in each person's interest in gaining a degree of control over the societal conditions of their life and concerns. Hence, Holzkamp argues for a "reason discourse" (in contrast to the still widespread "conditioning discourse") as the scientific language adequate to the task of formulating psychological theory and methods. Since reasons for actions must always be expressed in the "first person" mode, as "my reasons" from each individual subject's standpoint, the view of others from the external standpoint (as adopted in the conditioning discourse) has to be replaced by the standpoint of the human subject as the (necessarily always limited, partial and particular) scientific standpoint of psychological research.

This tradition has provided the basis for a *psychology of technology from the standpoint of the human subject* (Schraube, 2013). Such an approach is developed in contrast to an anthropocentric vision of science and technology. Similar to other psychological and STS standpoint theories (Harding, 2004; Martín-Baró, 1994) it tries to overcome supremacist, subjugating, top-down approaches in the production of scientific knowledge toward a bottom-up perspective rooted in the everyday problems, dilemmas, and concerns of people as well as the responsibility of humans for the self-created societal and technological relations. A central focus here is on the dialectics of technology and a critical analysis of both the potentiality as well as the power, constraints, one-sidedness, and discrepancies materialized in technological artifacts and systems. In this context, research has examined various arenas of everyday life such as, for instance, the significance of material artifacts in young children's conduct of everyday life (Chimirri, 2014), the automatization of work (Axel, 2002; Langemeyer, 2015), the digitalization of educational practice and learning (Schraube & Marvakis, 2016; Sørensen, 2009), questions of design and everyday practice (Costall & Dreier, 2006), or technoscience and the politics of experimental practice (Papadopoulos, 2018).

The subject-scientific concept *conduct of everyday life* (Holzkamp, 2013b; Schraube & Højholt, 2016) examines how, from their own situatedness, people relate to and act with technological artifacts, and seek, through those actions, to produce and sustain particular aspects of the world while changing others. This concept allows the psychology of technology to investigate how the subjective organization of everyday life and the socio-material situatedness of human agency are connected beyond the multiplicity of contexts of everyday action. This, in turn, makes it possible to seriously consider the complexity, dynamism and processual nature of the human relationship to technology.

The conduct of everyday life implies radically situating the analytic reflexivity of the psychology of technology simply because researchers are themselves actors within the relations under investigation. For this reason, such research employs collective and participatory methods. Here, then, the researchers are themselves regarded as part of their own psychological analysis of technology. Since such analyses also allow co-researchers to question each person's relations to the world, each participant researcher can negotiate, democratically (Nissen, 2012) and *teleogenetically* (Chimirri, 2015a), their own ideas and methods with the others, i.e., with regard to the impact which a (temporary) collective can hope for their ideas and methods to have on future societal action. As a result, the empirical researchers actively participate in the technological actions and practices of others, co-researching them through own experiences, exchanging their ideas and views on their shared as well as differing notions of technology and its objectives, taking up the contradictions that emerge, critiquing them together with the co-researchers, and locating them in their socio-historic context. The aim here is to achieve a generalized, but non-determining, understanding of how and why certain material artifacts ought to be kept and others changed and, in this way, shape a more democratic negotiation of future collective possibilities for action (Chimirri, 2019, 2015b). Partial and particular perspectives, including the immediacy and radical situatedness of everyday life experienced, can thus be temporarily transgressed, put at a critical analytical distance together. The standpoint of the human subject thus becomes generalized across more-than-human perspectives, and at the same time actively invites for getting critically inquired into and troubled by other human forces acting from within highly different life circumstances.

While dialectical practice approaches to the psychology of technology and feminist-posthumanist STS may draw on different philosophical foundations, both share a similar ethical and in consequence also onto-epistemological and political commitment. In her discussion of Barad's, Haraway's and others' work, cultural theorist Kathrin Thiele (2014) terms this commitment an *ethos of diffraction* that implies an *affirmative politics of difference(s)*:

> Affirmation (worthy of its name) practiced: Affirming that there will never be an innocent starting point for any ethico-political quest, because 'we' are always/already entangled with-in everything; and yet that this primary implicatedness is not bound to melancholy or resignation, which for too long has been preventing us to think-practice difference(s) that really might make a difference. (Thiele, 2014, p. 213)

Thiele argues for this ethos to transgress the post/humanist binary in theorizing ethics. In our eyes, this represents a central element that a dialectical practice psychology of technology is striving for: the particularity, partiality and the ensuing contradictoriness of knowledge demands us to put our respective knowledges to the test of difference, to have them explicitly challenged and troubled by more-than-human others. At the same time, however, it requires "staying with the trouble", as Haraway (2016) suggested: to take this troubling as a point of collective departure for affirmatively partaking in the critical inquiry and technological generalization of the internal, always already contradictory human–technology–world relatedness (in dialectical terms) or intra-active entangledness (in posthumanist terms). This ethico-onto-epistemo-methodological alliance across the post/humanist binary and differently attuned analytical apparatuses (with a conceptual emphasis on subjectivity and intra-acting more-than-human forces respectively), we hope, might help to substantiate a psychology of technology engaged in overcoming the Promethean gap identified by Günther Anders more than half a century ago.

## References

Anders, G. (1992). Die Antiquiertheit des Proletariats. *Forum, 39*(462–464), 7–11.

Anders, G. (2018a). *Die Antiquiertheit des Menschen. Band 1. Über die Seele im Zeitalter der zweiten industriellen Revolution*. München: Beck (Original work published 1956).

Anders, G. (2018b). *Die Antiquiertheit des Menschen. Band 2. Über die Zerstörung des Lebens im Zeitalter der dritten industriellen Revolution.* München: Beck (Original work published 1980).

Axel, E. (2002). *Regulation as productive tool use: Participatory observation in the control room of a district heating system*. Frederiksberg: Roskilde University Press.

Bakan, D. (1967). *On method: Toward a reconstruction of psychological investigation*. San Francisco: Jossey-Bass.

Bang, J. (2012). Aesthetic play: The meaning of music technologies for children's development. *Journal für Psychologie, 20*(1). https://www.journal-fuer-psychologie.de/index.php/jfp/article/view/114. Accessed 5 March 2018.

Barad, K. (2007). *Meeting the universe halfway: Quantum physics and the entanglement of matter and meaning*. Durham, NC: Duke University Press.

Barad, K. (2014). Diffracting diffraction: Cutting together-apart. *Parallax, 20*(3), 168–187.

Baucom, I., & Omelsky, M. (2017). Knowledge in the age of climate change. *The South Atlantic Quarterly, 116*(1), 1–18.

Becker-Schmidt, R. (1989). Technik und Sozialisation. Sozialpsychologische und kulturanthropologische Notizen zur Technikentwicklung. In D. Becker, R. Becker-Schmidt, G.-A. Knapp, & A. Wacker (Eds.), *Zeitbilder der Technik. Essays zur Geschichte von Arbeit und Technologie* (pp. 17–74). Bonn: Dietz.

Bozalek, V., & Zembylas, M. (2017). Diffraction or reflection? Sketching the contours of two methodologies in educational research. *International Journal of Qualitative Studies in Education, 30*(2), 111–127.

Braidotti, R. (2017). Critical posthuman knowledges. *South Atlantic Quarterly, 116*(1), 83–96.

Brinkmann, S. (2012). *Qualitative inquiry in everyday life*. London: Sage.

Bungard, W., & Schultz-Gambard, J. (1988). Technikbewertung: Versäumnisse und Möglichkeiten der Psychologie. In W. Bungard & H. Lenk (Eds.), *Technikbewertung. Philosophische und psychologische Perspektiven* (pp. 157–182). Frankfurt/M.: Suhrkamp.

Chimirri, N. A. (2014). *Investigating media artifacts with children: Conceptualizing a collaborative exploration of the sociomaterial conduct of everyday life*. Roskilde: Roskilde University.

Chimirri, N. A. (2015a). Moving as conducting everyday life: Experiencing and imagining for teleogenetic collaboration. In B. Wagoner, N. Chaudhary, & P. Hviid (Eds.), *Integrating experiences: Body and mind moving between contexts* (pp. 179–197). Charlotte, NC: Information Age Publishing.

Chimirri, N. A. (2015b). Designing psychological co-research of emancipatory-technical relevance across age thresholds. *Outlines, 16*(2), 26–51.

Chimirri, N. A. (2019). Specifying the ethics of teleogenetic collaboration for research with children and other vital forces: A critical inquiry into dialectical praxis psychology via posthumanist theorizing. *Human Arenas*. Advance online publication. https://doi.org/10.1007/s42087-019-00069-7.

Cole, M. (1996). *Cultural psychology: A once and future discipline*. Cambridge, MA: Harvard University Press.

Costall, A., & Dreier, O. (Eds.). (2006). *Doing things with things: The design and use of everyday objects*. Aldershot: Ashgate.

Dreier, O. (2007). Generality and particularity of knowledge. In V. van Deventer, M. Terre Blanche, E. Fourie, & P. Segalo (Eds.), *Citizen city: Between constructing agent and constructed agency* (pp. 188–196). Concord: Captus.

Engeström, Y., Miettinen, R., & Punamäki, R.-L. (Eds.). (1999). *Perspectives on activity theory*. Cambridge: Cambridge University Press.

Feenberg, A. (2017). Critical theory of technology and STS. *Thesis Eleven, 138*(1), 3–12.
Gergen, K. J. (2000). *The saturated self: Dilemmas of identity in contemporary life*. New York: Basic Books.
Gibson, J. J. (1986). *The ecological approach to visual perception*. New York, NY: Psychology Press (Original work published 1979).
Gordo-López, Á. J., & Parker, I. (Eds.). (1999). *Cyberpsychology*. London: Routledge.
Haraway, D. J. (1991). *Simians, cyborgs, and women: The reinvention of nature*. New York: Routledge.
Haraway, D. J. (1997). *Modest_Witness@Second_Millenium: FemaleMan©_Meets_OncoMouse™*. London: Routledge.
Haraway, D. (2016). *Staying with the trouble: Making kin in the Chthulucene*. Durham: Duke University Press.
Harding, S. (Ed.). (2004). *The feminist standpoint theory reader: Intellectual and political controversies*. New York: Routledge.
Hasse, C. (2015). *An anthropology of learning: On nested frictions and cultural engagements in organisations*. Dordrecht: Springer.
Hess, D. J. (2000). Ethnography and the development of science and technology studies. In P. Atkinson, A. Coffey, S. Delamont, L. Lofland, & J. Lofland (Eds.), *Handbook of ethnography* (pp. 234–245). Thousand Oaks: Sage.
Højgaard, L., & Søndergaard, D. M. (2011). Theorizing the complexities of discursive and material subjectivity: Agential realism and poststructural analyses. *Theory & Psychology, 21*(3), 338–354.
Holzkamp, K. (1983). *Grundlegung der Psychologie*. Frankfurt/M.: Campus.
Holzkamp, K. (2013a). The development of critical psychology as a subject science. In E. Schraube & U. Osterkamp (Eds.), *Psychology from the standpoint of the subject: Selected writings of Klaus Holzkamp* (A. Boreham & U. Osterkamp, Trans.) (pp. 28–45). Basingstoke: Palgrave Macmillan.
Holzkamp, K. (2013b). Psychology: Social self-understanding on the reasons for action in the conduct of everyday life. In E. Schraube & U. Osterkamp (Eds.), *Psychology from the standpoint of the subject: Selected writings of Klaus Holzkamp* (A. Boreham & U. Osterkamp, Trans.) (pp. 233–341). Basingstoke: Palgrave Macmillan.
Ingold, T. (2013). *Making: Anthropology, archaeology, art and architecture*. London: Routledge.
Kaptelinin, V., & Nardi, B. (2006). *Acting with technology: Activity theory and interaction design*. Cambridge, MA: MIT Press.

Knorr-Cetina, K. (1981). *The manufacture of knowledge: An essay on the constructivist and contextual nature of science*. Oxford: Pergamon.
Kontopodis, M., Wulf, C., & Fichtner, B. (Eds.). (2011). *Children, development and education: Cultural, historical, anthropological perspectives*. Dordrecht: Springer.
Langemeyer, I. (2015). *Das Wissen der Achtsamkeit: Kooperative Kompetenz in komplexen Arbeitsprozessen*. Münster: Waxmann.
Langemeyer, I. (2019). Beyond the cyborg-metaphor: Psychology in times of smart systems. In K. O'Doherty, L. Osbeck, E. Schraube, & J. Yen (Eds.), *Psychological studies of science and technology*. London: Palgrave Macmillan.
Latour, B. (2005). *Reassembling the social: An introduction to actor-network theory*. New York: Oxford University Press.
Latour, B. (2017). *Facing Gaia: Eight lectures on the new climatic regime*. Cambridge: Polity Press.
Latour, B., & Woolgar, S. (1986). *Laboratory life: The social construction of scientific facts*. Princeton: Princeton University Press.
Law, J. (2007). Making a mess with method. In W. Outhwaite & S. P. Turner (Eds.), *The SAGE handbook of social science methodology* (pp. 595–606). London: Sage.
Leontyev, A. N. (1981). *Problems of the development of the mind*. Moscow: Progress Publishers.
Luria, A. R. (1928). The problem of the cultural behavior of the child. *Journal of Genetic Psychology, 35*, 493–508.
Martín-Baró, I. (1994). *Writings for a liberation psychology*. Cambridge, MA: Harvard University Press.
Marvakis, A. (2013, May 3–7). *Re-reading Marx for psychology, e.g. alienation I*. Paper presented at the 15th Biennial Conference of the International Society for Theoretical Psychology, Santiago, Chile.
Mol, A. (2008). I eat an apple: On theorizing subjectivities. *Subjectivity, 22*, 28–37.
Nissen, M. (2012). *The subjectivity of participation: Articulating social work practice with youth in Copenhagen*. Basingstoke: Palgrave Macmillan.
Ollman, B. (2003). *Dance of the dialectic: Steps in Marx's method*. Urbana: University of Illinois Press.
Ollman, B. (2015). Marxism and the philosophy of internal relations; or, how to replace the mysterious 'paradox' with 'contradictions' that can be studied and resolved. *Capital & Class, 39*(1), 7–23.
Papadopoulos, D. (2018). *Experimental practice: Technoscience, alterontologies, and more-than-social movements*. Durham, NC: Duke University Press.

Puig de la Bellacasa, M. (2017). *Matters of care: Speculative ethics in more than human worlds*. Minneapolis: University of Minnesota Press.
Ripple, W. J., Wolf, C., Newsome, T. M., Galetti, M., Alamgir, M., Crist, E., ... Laurance, W. F. (2017). World scientists' warning to humanity: A second notice. *BioScience, 67*(12), 1026–1028.
Scarry, E. (1985). *The body in pain: The making and unmaking of the world*. New York: Oxford University Press.
Schatzki, T. R., Knorr Cetina, K., & von Savigny, E. (Eds.). (2001). *The practice turn in contemporary theory*. New York: Routledge.
Schraube, E. (2009). Technology as materialized action and its ambivalences. *Theory & Psychology, 19*(2), 296–312.
Schraube, E. (2013). First-person perspective and sociomaterial decentering: Studying technology from the standpoint of the subject. *Subjectivity, 6*(1), 12–32.
Schraube, E. (2019). Technology and the practice of everyday living. In H. J. Stam & H. Looren de Jong (Eds.), *The SAGE handbook of theoretical psychology*. London: Sage.
Schraube, E., & Højholt, C. (Eds.). (2016). *Psychology and the conduct of everyday life*. London: Routledge.
Schraube, E., & Marvakis, A. (2016). Frozen fluidity: Digital technologies and the transformation of students learning and conduct of everyday life. In E. Schraube & C. Højholt (Eds.), *Psychology and the conduct of everyday life* (pp. 205–225). London: Routledge.
Seghal, M. (2014). Diffractive propositions: Reading Alfred North Whitehead with Donna Haraway and Karen Barad. *Parallax, 20*(3), 188–201.
Sørensen, E. (2009). *The materiality of learning*. Cambridge: Cambridge University Press.
Stengers, I. (2015). *In catastrophic times: Resisting the coming barbarism*. Lüneburg: Meson Press.
Stetsenko, A. (2005). Activity as object-related: Resolving the dichotomy of individual and collective planes of activity. *Mind, Culture, and Activity, 12*(1), 70–88.
Teo, T. (2009). Philosophical concerns in critical psychology. In D. Fox, I. Prilleltensky, & S. Austin (Eds.), *Critical psychology: An introduction* (pp. 36–53). London: Sage.
Teo, T. (2017). From psychological science to the psychological humanities: Building a general theory of subjectivity. *Review of General Psychology, 21*(4), 281–291.

Thiele, K. (2014). Ethos of diffraction: New paradigms for a (post) humanist ethics. *Parallax, 20*(3), 202–216.
Turkle, S. (1984). *The second self: Computers and the human spirit*. New York: Simon & Schuster.
Turkle, S. (2015). *Reclaiming conversation: The power of talk in a digital age*. New York: Penguin Press.
Walkerdine, V. (1997). Postmodernity, subjectivity and the media. In T. Ibanez & L. Iniguez (Eds.), *Critical social psychology* (pp. 169–177). London: Sage.
Wartofsky, M. W. (1979). *Models: Representation and the scientific understanding*. Dordrecht: Reidel.
Winner, L. (1989). *The whale and the reactor: A search for limits in an age of high technology*. Chicago: University of Chicago Press.
Winner, L. (1996). The gloves come off: Shattered alliances in science and technology studies. In A. Ross (Ed.), *Science wars* (pp. 102–113). Durham: Duke University Press.
Zuboff, S. (2019). *The age of surveillance capitalism: The fight for a human future at the new frontier of power*. New York: Public Affairs.

# 4

# Neuroscience and the New Psychologies: Epistemological First Aid

Henderikus J. Stam

In an era of neuroscience, what can psychological studies of science and technology tell us? Medical models dominate, research is intensely focused on a few technologies, and the new findings appear rapidly while being fragmented and discontinuous. What I hope to show in this chapter is that our response to the neurosciences need not require wholesale adaptations of contemporary psychology but can be treated like a psychological problem in its own right. Psychological studies of the neurosciences can be directed appropriately to contain some of the wide diffusion of claims and counterclaims that are populating the literature.

"Few scientific areas have captured the popular imagination more or seem to relate more cogently to human concerns than the neural sciences." So began a book by Daniel Robinson, but surprisingly those words were written in 1973. He continued, "As if the intrinsic glamor

H. J. Stam (✉)
Department of Psychology, University of Calgary,
Calgary, AB, Canada
e-mail: stam@ucalgary.ca





K. C. O'Doherty et al. (eds.), *Psychological Studies of Science and Technology*, Palgrave Studies in the Theory and History of Psychology,
https://doi.org/10.1007/978-3-030-25308-0_4

of the field were not enough, the full arsenal of modern press-agentry has accompanied recent advances. As a result, the layman is often left believing that 'they' are at the threshold of discoveries that may irreversibly alter the destiny of his species" (Robinson, 1973, p. 1).

Reporting on the advances of the neurosciences in topics traditionally psychological is no longer news and the developments of these sciences have remade psychology academically, technically, practically as well as in the popular imagination. The refinement of electroencephalography measures along with the addition of such tools as positron emission tomography, computerized tomography, optical tomography, and functional magnetic resonance imaging has changed the nature of speculation about the brain just as it has made these more visible. That visibility is nonetheless dependent on sophisticated statistical and constructive mathematical and computerized processes.

As if continually restating Robinson's argument, for the better part of two decades these influences have become an ever more present reality in most psychology departments in the industrialized world (or for the sake of historical accuracy now frequently referred to as "post-industrial societies" [Bell, 1974]), pushed along by a formidable interdisciplinary, multipronged, and richly funded matrix of research, tools, and practices whose very pronouncements promise the revelations of ever greater truths about ourselves and our brains.

To make the case bluntly, witness a recent issue of the *Harvard Business Review*. Three authors enthusiastically promote applied neuroscience for the purposes of generating new understandings of customers. They argue that "applied neuroscience is used primarily during one of two points in a new project — either at the onset while *defining* the business problem, or later in the cycle while *seeking* new solutions for users" (Furr, Nel, & Ramsøy, 2019). What exactly is meant by all this is unclear, but the enthusiasm is not.

At the same time, the neuroskeptics and critics in psychology have not held their fire, and critiques of the influence of the neurosciences on psychology are legion. Among the skeptics are those who note the failure of much neuroscience research to replicate, the lack of integration of the neurosciences within psychology, and the piecemeal nature of much of neuropsychological research (e.g., Choudhury & Slaby, 2012;

Gergen, 2010). Much neuroscience research is also profoundly inductivist and has been subject to critical philosophical analysis (e.g., Bennet & Hacker, 2003; Bennett, Dennett, Hacker, Searle, & Robinson, 2007).

Critics who have worried about the reductionist intent of the neurosciences see this as a continuation of a radical or elminativist reductionist worry that has been part of discussions in the philosophy of psychology for some 70 years (e.g., Margolis, 1984). And the arguments against eliminativism that were posted then are equally relevant now, except that most neuroscience research in psychology is not explicitly elminativist. Instead it hedges its bets; it seeks mere associations between certain kinds of fragmentary pictures of brain states and psychological events, dispositions, or actions. Even the Libet experiments which presumably demonstrate that a "readiness potential" initiates the free will to act are now controversial, to say the least (Radder & Meynen, 2012).

Other critiques are varieties of the mereological fallacy so clearly developed by Bennett and Hacker in 2003. A mereological fallacy is one in which the powers and activities of human persons are attributed to brains or parts of brains. Examples abound, such as found in popularizations of brain research of the sort that showed up in a recent issue of the *Atlantic*. This was a story that claimed that neuroscience research had shown by way of TMS stimulation that powerful people were impaired in a neural process called "mirroring" or the ability to feel empathy with the less powerful (Useem, 2017).

Other critics have noted the limitations of neural processes in explaining complex social activities. For example, Coey, Varlet, and Richardson (2012) have argued that understanding the context of social interactions requires understanding their "embodied-embedded" constraints. These and other authors have argued that the organization of human activity, particularly its self-organizing processes, requires something much more dynamic than a neural account. Others have noted that most forms of psychopathology are not single entities at the level of the brain. For example, Roiser (2015) has noted that "despite the statistical evidence for group differences, even the most robust brain-imaging abnormality in depression (reduced sgACC [Subgenual anterior cingulate cortex (sgACC)] volume) cannot differentiate between depressed and never-depressed individuals reliably."

Feminists have argued that this too is another science caught up in gendered cultural symbolism and power relationships (e.g., Schmitz & Höppner, 2014). Jordan-Young and Rumiati (2012) argue that the neurosciences make gender structures appear "natural and inevitable." Indeed many of the arguments criticizing traditional categories of sex and gender in psychology are applicable to the neurosciences as well.

All of these critiques have their place. A reductionist neuroscientific language after all cannot replace the reporting role of ordinary language, that is the language of intentions, semantics, and sentience. If it could, it would have to be as contextually sensitive as ordinary language and the neurological language would simply become as flexible as ordinary language is now, losing its scientific qualifications. Not that reductionist accounts are not useful, they are important guideposts to understanding the brain *qua* brain. It's just that they can't replace ordinary language accounts of human action.

The results of these critiques are a kind of intellectual paralysis combined with a great deal of piecemeal research activity. In commenting on the state of the discipline vis a vis neuroscience, psychologists such as Scott Lilienfeld (a vocal proponent of a "clinical science model") and colleagues recently wrote that "brain science *is* an integral component of psychology" (Schwartz, Lilienfeld, Meca, & Sauvigné, 2016, p. 65) Following this less-than-profound claim, he and his coauthors argue, "social-science concerns are essential as well, and that psychology must maintain a focus on such concerns as it continues to expand into natural science domains" (p. 65). After several more such non-informative claims, the authors state that "Clearly, neuroscience has brought an enormous wealth of data and knowledge to the field of psychology. It has helped psychologists to begin to demystify the brain, to learn how the brain develops and adapts, and to better identify the links between brain functions and psychological processes." Although worried about the hiring trends in psychology departments which have focused, at least in North America, on bringing ever more neuroscientifically oriented researchers into psychology, Lilienfeld and colleagues argue that psychology should "aim to be inclusive." What is clearly missing is a vision of just what *psychology* is about and just what makes it an independent discipline.

## Psychology, Technoscience, and STS

All of this hand-wringing is more like a closing of the barn door after the horse has bolted. And calls for "balanced" approaches are reminiscent of the heyday of behaviorism when voices of reason called for more balanced approaches to the discipline but the allure of a straightforward mechanical solution to human psychology in the form of the study of behavior was too strong. I want to argue from another perspective today that the inevitable advances in neuroscience do not mean the end of psychological questions, merely their reformulation and only then in some very few instances. There are more interesting ways to engage the neurosciences. Indeed, I will argue for the need of some *epistemological first aid* in doing so. In discussing this possibility it will be important to note that the neurosciences are typically viewed as a one way street—they influence how and what is important in psychology. But the other direction is just as, if not more important for it is our psychological take on the world that deeply influences what we take to be important in the neurosciences. And what we take to be important psychologically is prima facie what counts as psychological in a shared world of cultural norms and mores.

In what follows I will elaborate on a conceptual tool that I think might be useful, something I have called "epistemological first aid" borrowing from Solymosi (2014) who has modified Dennett's (1988) conception of "ethical first aid." This is premised on the notion, discussed below, that neuroscience can be an assistant to clarifying what we know about ourselves and our capacities rather than a force that diminishes our traditional self-understanding in favor of some physio-neuro-reductionist account. In other words, rather than treating our attributes, experiences, and intentions as the outcome of some neurological process, we take those neurological processes as subsidiary processes that may, on occasion, clarify our psychological life. It is the latter that is primary, whatever neuroscience can tell us about ourselves is beholden to the primacy of experience. I hope this will become clear as I go along.

I will begin with a stalwart critic of psychology, Nikolas Rose. We know Rose largely as the critic of psychology and all "psy" disciplines, who argues that those disciplines are engaged in practices of governmentality, after Foucault, referring in particular to the creation of subjectivities through the organized practices of a society (e.g., Rose, O'Malley, & Valverde, 2009). But in his recent book with Joelle Abi-Rached (2013) entitled "Neuro," the authors argue that "despite their apparent contradictions, neurobiological research emphasizing the role of nonconscious neural processes and habits in our decisions and actions can-and does—happily coexist with longstanding ideas about choice, responsibility, and consciousness that are so crucial to contemporary advanced liberal societies" (Rose & Abi-Rached, 2013, p. 21). So for Abi-Rached and Rose, neuroscience has not removed from us our responsibility to be actors whose fates are not captured only by processes that occur outside awareness in our brains, but also has not lessened the requirement that we govern those forces through an endless process of self-discipline. In other words, neurosciences support the status quo. It is Abi-Rached and Rose's optimistic view that the eminent sociality of human existence requires that a neuro-reductive language will have to address "questions of complexity and emergence, and to locate neural processes firmly in the dimensions of time, development, and transactions within a milieu" (Rose & Abi-Rached, 2013, p. 23). In other words, the picture of our brain as plastic and ultimately social is a revisionist one that can be used for multiple ends and will save psychology from being overwhelmed by the neurosciences. But here I find Abi-Rached and Rose curiously concerned with balance, almost in the same vein as Lilienfeld and colleagues. One might even call their position a conservative one in so far as it does not challenge the neuroscientific imperative to recreate psychology in its image. Abi-Rached and Rose seemed more preoccupied with the finer details of the neuroscience literature than its larger impact. Granted, given Rose's broader critique of the "psy" disciplines, perhaps he is not particularly concerned with the disciplinary anxiety of whether the neurosciences swallow psychology or vice versa. However, he and Abi-Rached are strangely quiet on the broader implications of the neurosciences for our expanding notions of subjectivity.

Recently I argued that an alternative view, drawing on the work of Kevin Moore and Maarten Derksen, among others, might be more productive (Stam, 2015). Using Andy Clark's metaphor of the brain as an instrument is one way of moving beyond a neuro-reductive account argues Derksen, because we are both identified with our brains and treat it as something external to us. The brain as instrument is an attempt to steer between a version of personhood that makes us neither the passive bystanders of what happens in "our brain" nor does it make us able to "use" the brain just as we will. Technologies are societies made durable, in Latour's (1990) terms.

I want to revisit this conception since it is neither obvious nor amenable to the kinds of popular generalizations of brain science found in the media and sometimes in our psychology texts. I also want to revisit it because I felt I did not go far enough.

In my original argument, I used Latour's example of the development of a "nose" for perfume (or I could also have used the development of a palate for wine). In developing a "nose" for perfume, Latour argues, the neophyte must learn to differentiate among many odors. This is accompanied by the articulation of the contents of different odors typically after lengthy practice. The exact, precise chemical foundation of an odor is not important since this is a matter of *accuracy of reference*. As Latour (2004, pp. 210–211) put it, "the decisive advantage of articulation over accuracy of reference is that there is no end to articulation whereas there is an end to accuracy."

But transposed to the brain sciences this doesn't quite work the same way. Certainly what a brain is capable of, how it makes a difference in life, what it allows us to do, and so on, has immediate social relevance. And the greater our capacity for articulation, the greater our capacity for insight into its failures and its vagaries. To know for example that there are pathways in the dentate gyrus that erase memories is to confirm that our memories are labile and conforms to a host of discourses on eyewitness testimony and the like (Madroñal et al., 2016).

Articulating this is not the same as an articulation of odors in learning to develop a nose for perfume or tastes that develop a palate for wine. While the latter consist of fine-tuning skills that have immediate social import, our language of brain states requires further elaboration of their place in a social context that values such talk. Furthermore, is its value in part, a case of revealing our bodies to us the way that X-rays are revealing? When we view an X-ray of a skull with a fracture, the gray image of a head is taken as a practical illumination of what is under our skin. It is us and it is not us at the same time. When an X-ray of our broken wrist is displayed, we understand that this too is a part of us—both as object and as problem. As a technology it is both distancing and revealing. It looks like something other than us, while we recognize that it also reveals who we are and is made possible by a vast network of medical practices that have shaped bodily existence in the twentieth century and beyond.

But brain scans are different again from X-rays. As Joseph Dumit, among others, already argued in 2004, they are historical, experimental, popular, mediatized, and clinical objects (de Rijcke & Beaulieu, 2007). Dumit argued that brain scans provide powerful semiotics of what is normal (Chapter 1). He says,

> We might call the acts that concern our brains and our bodies that we derive from received-facts of science and medicine the *objective-self*. The objective-self consists of our taken-for-granted notions, theories, and tendencies regarding human bodies, brains, and kinds considered as objective, referential, extrinsic, and objects of science and medicine. That we "know" we have a brain and that the brain is necessary for our self is one aspect of our objective-self. We can immediately see that each of our objective-selves is, in general, dependent on how we came to know them. Furthermore, objective-selves are not finished but incomplete and in process. With received-facts, we fashion and refashion our objective-selves. Thus it is we come to know certain facts about our body as endangered by poisons like saccharine, our brains as having a "reading circuit," and our fellow human beings as mentally ill or sane or borderline. (p. 7)

The ambiguity of these images, their use in scientific publications as well as popular culture, means that there is no final arbiter of just "how" we are to read them. Hence there is no final story to be told about what our brains are, they are works in progress and hence can be used to various social ends. As de Rijcke and Beaulieu (2007) note, "Images are neither self-explanatory nor transparent, but rather partake in a specific visual culture. If they are to serve as bridges (e.g. to popularize scientific results), then they are only effective insofar as cultural conventions are shared" (p. 733).

Neuroscience is of course about much more than brain scans. A complex set of interrelated disciplines working on, that is experimenting, observing, analyzing nonhuman animal and human brains and nervous systems, do not produce a coherent and easily narrativized story about the brain. Instead, they work to complicate matters continually. Think, for example, of the gut-brain axis, said to be one of the "new frontiers of neuroscience" (Foster, 2013). Microbiota in the gut, also called the "second brain," are purported to have a relationship to mood disorders. While no clear linear connection has yet emerged, there are sufficient grounds for neuroscientists to continue this line of investigations. Suddenly bacteria and other microorganisms in the gut form part of the investigations of neuroscientists. A recent article in *Biology Direct* presented the hypothesis that certain religious rituals were created by our microbiome to encourage microbial transmission. The authors said, "We hypothesize that certain aspects of religious behavior observed in human society could be influenced by microbial host control and that the transmission of some religious rituals could be regarded as the simultaneous transmission of both ideas (memes) and parasitic organisms" (Panchin, Tuzhikov, & Panchin, 2014). In other words, next time you go to the temple or church, thank your microbiome for sending you there. These claims stretch the evidence to a point of the absurd, but the fact that the article was published in a serious journal means that we are dealing with multiple versions of human persons or "objective selves" in the neurosciences. Surrounded as we are by the constant reminders that the neurosciences are here to explain us to ourselves, we are not getting a very clear story. In other words technologies of the brain are always under construction—neither passive tools nor deterministic but dependent on a large range of actors and technical devices.

What are we getting then when we engage in brain talk? Not the kind of skill that one gets when one develops a nose for perfume or a palate for wine. But we are confronted with a technology that is becoming a part of our psychological and popular discourse of selves. Like psychoanalytic terms, such as the "unconscious," and behavioral terms, such as "learning curves," or cognitive terms, such as "representations" we can expect that neuroscientific terminology and concepts will increasingly inhabit our daily speech and self-understandings.

Such self-understandings are like Gigerenzer's consideration of tools-to-theories (Gigerenzer, 1991), as we develop familiarity with a vocabulary and the tools that produce it, the understandings that follow seem logical, necessary, and even preordained. Resistance is at once futile and necessary. Futile in the same way that arguing, for example, that there may be no such thing as a (Freudian) unconscious will not do much to change the way that notion has infiltrated common understandings. But there is no end to the way in which our conception of the unconscious has been reconceptualized, both in ordinary language and in psychology and psychiatry, and now neuroscience and neuropsychoanalysis. We can only expect that neurological terms will infiltrate ordinary language in the same way, just as the word brain itself is already a repository of surplus meaning. Reflexive discourses of the brain are likely to change our intuitions about ourselves.

## Law as Model?

The influence of the neurosciences on the law could serve as an illustration. The law after all is a repository as well as enforcement of moral codes, even if they don't overlap in any necessary way. And common law is a repository of hundreds of years of judgments concerning those codes. Neuroscience, like any science potentially, can affect legal cases wherever that science is relevant. But neuroscience has a unique role in so far as it will lead the legal system to question key notions of responsibility that are central to determinations of guilt or innocence. As Greene and Cohen (2004, p. 1775) argue,

> ......neuroscience will probably have a transformative effect on the law, despite the fact that existing legal doctrine can, in principle, accommodate whatever neuroscience will tell us. New neuroscience will change the law, not by undermining its current assumptions, but by transforming people's moral intuitions about free will and responsibility. This change in moral outlook will result not from the discovery of crucial new facts or clever new arguments, but from a new appreciation of old arguments, bolstered by vivid new illustrations provided by cognitive neuroscience.

As Roger Smith (2010) has so beautifully argued, what neuroscience has to say about "will" is largely in the order of propaganda, and that the neurosciences have nothing to say about the will since the will is a kind of collective agency, not an individual feeling. That is, what makes "will" possible is the communal structures of agreed upon existing life. In the same way, psychology may be able to accommodate "whatever neuroscience will tell us" but it affects so many aspects of what it is to be human that psychology will be tempted to shift conceptions of personhood in the process.

Take for example, Jonathan Haidt's (2012) claims about the so-called "automaticity revolution." To quote Haidt (2012), "within the first quarter of a second, we react to people's faces, we react to words, we react to propositions, and then reasoning is much slower." Regardless of the robustness of these findings they change nothing about our capacity to reason or to consider arguments. That people have preferences, biases, prejudicial beliefs, and so on, are not news either. But the so-called automaticity revolution heralds a version of public accounting that takes the mundane prejudices of everyday life and valorizes them by giving them a neural basis. And here one must ask the obvious: isn't the purpose of years of education (our so-called long childhood), of most cultural norms and mores, and of the social constraints present in everyday life just to overcome what we know intuitively to be those immediate prejudices and judgments. What neuroscience and self-styled moral psychologists like Haidt tell us is that this is new knowledge. What a knowledge of history and culture teaches us is that we already know this in our common cultural heritage and that the social organization of human life has over the centuries built into our customs the very defenses needed to refute those judgments. In this case, it is not that the neuroscience is not important or interesting, it's just that a historical view will temper the "wow" effect of these so-called discoveries and their naïve interpretations in the hands of psychologists. Historically, the neurosciences are one more technology for not only self-understanding but for the social management of self-understanding. In their proper historical context, they take their place among the multiple technologies of the self that already populate our existence. It is precisely at this juncture that it is helpful to think of epistemological first aid.

## Scientific vs. Manifest Images of Humanity

The quandary between technologies that threaten our common self-understanding has been characterized by philosophers as the distinction between science and commonsense (Solymosi, 2014), or as Sellars has noted between the scientific and manifest images of humanity. The manifest is what we take traditionally to be the governing conception of our selves, namely that we have something like a free will, we are capable of rational decisions, we have a personal history/identity we can claim and so on. Science on the other hand often denies or confounds this; our will is not free or not as free as we would like to think, our decision-making is influenced by far more subtle and not so subtle features of our bodies and the world and our personal history is largely one derived from a highly egocentric and distorted view of the world. For Solymosi (2014) the philosophical problem has been conceptualized as one largely of the *reconciliation* of these two views whereas we would be better served instead to consider Dewey's view (argues Solymosi) of the *reconstruction* that is inherent in our relationship to science. In particular, Solymosi believes that Dewey's formulation consists in not as seeing science as a threat to our ideals but rather to see it as a way of achieving these ideals. This is so because "in understanding the operations of nature we are able to guide our further behavior in ways more amenable to amelioration of our perceived problems and to the consummation of our ideals" (Solymosi, 2014, p. 297).

Solymosi is particularly concerned with the reconstruction of morality and ethics in the light of neuroscience. To get there he recruits Daniel Dennett's notion of "moral first aid." In Dennett's (1988) view, not only are traditional conceptions of ethics impractical but moral experience is not detached from having to make judgments in the here and now. In order to do so, Dennett's (and by extension, Solymosi's) notion of the "moral first aid manual" is premised on the notion that we must make decisions in the way nautical manuals served early navigators. Without having to do the weighty calculations, a manual makes it possible to make decisions. Now, Solymosi tries to work out how this might work in the case of ethical decision-making, given that our knowledge is in flux and our reconstructive ethical capacities must make use of "goods, rights, and virtues in a variety of problematic situations" (p. 303). Ultimately, Solymosi argues,

> In reconstructing experience as the transaction of organism and environment, and in conceiving of science in Hickman's technological sense[1], ethical inquiry—moral first aid—is a matter of engineering new tools of climbing or navigating the moral landscape. It is also a matter of engineering new ways of altering the landscape itself through our technical interactions with it. Depending on the platforms or cranes on which we stand, we may find new insights into particular ethical problems, such as those surrounding the disorders of consciousness... (p. 311)

I have not done justice to the entirety of Solymosi's argument here but suffice it to say that the question that interests me is its implications for psychology. The discipline's current desire to map its major domains in the language and tools of neuroscience requires a rethinking of the fundamental goal of psychology. Here it is that we require an *epistemological first aid*. This is not meant entirely tongue in cheek, for insofar as advances in psychology become part of our tradition of solving real problems; we have introduced new ways of *being* psychological. This is neither a foregone conclusion—it does not immediately upend the traditions of taken-for-granted notions of persons—nor does it invalidate other psychological accounts. For there is no single psychological account of persons to which we, as members of post-industrial societies, adhere. Indeed, psychology competes with many other forms of knowledge production and cultural traditions for the right to define what constitutes human nature. In its adherence to a form of scientific practice however it claims to extend its epistemic authority beyond that granted to many other concerns. But it's precisely in the competition of accounts that we find possibilities of being that require of us to

---

[1]Solymosi is here referring to Larry Hickman, who argues that technoscience is the more appropriate term for what is nominally called "science" since it is in fact a branch of technology that involves the use of tools and artifacts that are employed on raw materials to solve problems (see Hickman, 2001). Science does no more come before technology than it is separate from it. Techoscience is on this account continuous with human experience.

reformulate who we are. Some such reformulations will be failures, in the long run (take radical behaviorism as a case in point, if not, original formulations of psychoanalysis might do). But we have no obvious foresight into the long run, nor do scientific predictions hold much water, particularly those that have tried over the last 100 years to predict what psychology might become.

What neuroscientific accounts frequently imply however is that instead of multiple psychologies there is only one arising out of a single version of the brain. It is here that certain controversies arise and where epistemological first aid is most necessary. For whatever the neurosciences may yet portend, that neural responses *must* reflect a unified function of the brain is a highly unlikely scenario (Anderson, 2014). Anderson argues that in the neurosciences "the focus on local, linear correlations between brain activity and simple stimuli will never be by itself sufficient to capture the complexity of the brain and its interacting parts" (p. 10). Furthermore acknowledging this will require a rethinking of the vocabulary of cognition, "We will deeply rethink the vocabulary of cognition, ideally giving the brain a voice in the process. In discerning what the brain cares about, we will remember that it evolved to be an action-control system, specializing in managing the values of salient organism-environment relationships. Hence, many of the properties to which the brain is attuned will be action-relevant and relational" (p. 10).

What kind of psychological being we think we are as they are posited in the psychologies of the future are to be negotiated between science and life. We cannot take the findings of neuroscience as a given for how to live life, never mind how to understand the problems of life. Nonetheless we realize that the neurosciences may change elements of how we understand that life in the making or in its breaking. It is here that we are most dependent on epistemological first aid. Taking Solymosi (and Dewey) seriously, it is clear that forms of life are not fixed, they are ever shifting and shaping landscapes of bodies in coordinated action. How neuroscience fits into such a landscape is a question for the psychological sciences to discover as well as, possibly, resist. Psychology will remain independent of neuroscience if only because lived lives exist in social worlds, not of the brain's making.

That is, human life in its inherent sociality is both structured and constituted on the fly, a moment-by-moment recalibration of that life is the kind of story told, for example, in ethnomethodology. No amount of neuroscience can predict precisely the kind of social world that might emerge at any given moment in history.

One of the problems that Solymosi identified in Dennett's proposal for ethical first aid was the need for a foundational account of moral life and ethics. Dennett's naturalistic account can't provide one and hence "ethics and its product, morality, must be reconstructed in an experimental fashion" (p. 300). To solve this Solymosi sees the need for a pragmatic (in the Deweyan sense) ethical technology. For a psychology in need of an epistemological first aid, there is no need for first principles of the sort that ground epistemology since a technoscience is already thoroughly pragmatic. It is the community of knowledgeable users that ultimately determines just which epistemological questions will matter. This is going to be contested and assumes the free exchange of knowledge as a matter of course. For example, let's assume that there are good reasons for no longer using a notion like Hull's (1934) "habit family hierarchy" to try to understand learning just as we no longer employ phrenology to make judgments of character. At a certain point these become antiquarian, epistemological concepts. They do not make sense of life as lived and understood by the community of users (psychologists and most of the public), even if there are remnants of this knowledge in various corners of psychology. Even phrenology was not ruled out of court as "wrong" so much as misguided. For while it may have been mistaken about the brain, phrenologists practiced an early form of what we now see as clinical psychology, so their practices informed a future model of therapeutics. Epistemological first aid came to phrenology in the form of the "new psychology" of the late nineteenth century. It came to Hullian learning theory in the form of cognitive science. And each of those has subsequently been supplanted. Visions of human psychic life replace one another at fairly regular intervals—we can be confident that the neurosciences will be just one step in many to come. At some point, we require not a total overhaul but a first aid course to cure what ails our epistemic categories.

## Technoscience and the Human Project

It was John Dupré who noted some years ago that "ultimately, human evolution and human history are the same thing" (2001, p. 99). By this he meant that the vast changes in human behavior are as important as changes in human biology and the dividing line between them is fuzzy at best. Corrective lenses have, for generations, made it possible for people to be productive into old age, a phenomenon driven by literacy, itself a radical cultural change in human activity. Antibiotics, introduced in the middle of the twentieth century, have now kept alive a vast cohort of people who might otherwise not be here and hence we have, ipso facto, altered human genetics along the way. Such obvious if not banal observations are only meant to note that human life and the science that we produce are not just entwined but coexistent, or perhaps symbiotic. The products of technoscience are themselves embedded in the activities of human beings whose goals may be supportive or deleterious to human beings, but they are always part of our activities. In that sense, the neurosciences are not abstract pronouncements but always on the edge of what it is we can and might yet know about our brains and central nervous systems. That knowledge will not alter our fundamental need to solve the mundane problems of how we live life and how we continue to build social structures to support our human psychic life.

## Standing by the Bystander Effect

Perhaps I can clarify epistemological first aid by way of two examples. First, I would like to take an example from psychology, in this case the "bystander effect" (Darley & Latané, 1968). The bystander effect refers to the lack of assistance provided by bystanders in an emergency that is proportional to the number of witnesses present. The more witnesses, the less likely an individual is to assist.[2] Hortensius and

[2]This is actually controversial, like so many findings in social psychology. The results of multiple studies have complicated the original findings and there is no overall theoretical model to give an account of the findings that are reported (Fischer et al., 2011). However, the results of these

de Gelder (2014) demonstrated that when shown videos of an emergency, participants lying in an fMRI scanner will show neural activity commensurate with the number of bystanders present. The videos showed from zero to four bystanders walking past a person who faints. The authors conclude, "The left precentral and postcentral gyri and the left medial frontal gyrus showed a decrease in activity with the increase in group size. In contrast, regions involved in visual processing and attention showed an increase in activity with the increase in group size. We propose that these results support the conclusion that group size during an emergency already influences activity in brain regions sustaining preparation for action" (Hortensius & de Gelder, 2014, p. 56).

As a typical fMRI study, this is interesting, at least with respect to other fMRI studies concerned with localized and specific brain functions. It is an observation about activation in regions of the brain. My concern is the way in which such studies are treated in more popular contexts particularly in media for the "educated layperson" such as the following (McGrath, 2019), "Hortensius and de Gelder's breakthrough shows that, at least initially, there is no conscious decision *not* to intervene. Significantly, their work also indicates that the presence of others increases both the 3F response [fight-flight-freeze] and personal distress levels, indicating that trauma is amplified if experienced in a group." We need immediate first aid: we can quickly note that, first, there is no evidence in the study itself that "there is no conscious decision *not* to intervene" since the study was not capable of showing this. Second, the claim that "trauma is amplified if experienced in a group" was also not even remotely addressed by the study. Epistemologically it was a study of blood flow to certain regions of the brain under highly artificial conditions.

The complexities of bystander interventions prevent us from making the leap, however much the video clips mimic genuine emergencies.

studies have become embedded in the discourse of "bystanders" in North America. For example, my university offers training to faculty and students in "bystander intervention." I say this not to critique such training, which undoubtedly is useful for some ends, but to note how the notion of the inactive bystander has become a common trope.

But for the sake of argument, let's assume the videos accurately reflect what might happen in real life settings where a bystander confronts an incident requiring assistance. We might want to administer some more epistemological first aid before we go any further. An activated brain confirms that the person whose brain it is must consider new situations in multiple ways. Ultimately we confront a moral conundrum unique to urban life—do we intervene in the troubles of strangers? The many questions and contexts that might affect such an act still do not tell us what happens in the individual case. After all, it will be a person that intervenes or not, and that person will have to make a complex judgment based not only on their bodily state and their "preparation for action" but also on a series of moral and cognitive assessments that will likely be partial and incomplete. That others will influence such decisions is certainly likely, but just how and in what ways are not clear. Our modern urban existence makes all such judgments difficult, if not uncertain. Our knowledge of such judgments and the actions (or lack thereof) that follow are not captured by brain scans, even if we recognize that as a piece of technoscience, such scans add to our overall considerations of human action. Epistemologically we are not helped here by the brain scan even if neuroscience is helped in some obtuse way.

I fear that this example may seem trivial if not downright wrong-headed and the critique is perhaps obvious. Hortensius and de Gelder did not make the kinds of claims found in McGrath's summary of their research. So in the twenty-first century we have come a long way from a time when we might have tried to characterize the bystander effect in terms of character traits discovered through phrenology (although to my knowledge no one ever did so).[3] That technology is no longer accessible to a contemporary educated person, it would be a meaningless discussion and considered a waste of time. But that phrenology was practiced up to the time of WWII, however, demonstrates just how long a particularly persuasive discourse can survive. Functional magnetic

[3]It could be argued of course that there was no such thing as the "bystander effect" in the nineteenth century because the historical, urban conditions of alienation that led to this effect were not yet fully present. Hence no phrenological account could be provided.

resonance imaging is an important tool that we expect to provide us with certain results about blood flow in the brain. As the technology expands, we are carried away in its grip, which leads to discourses of science and brains quickly integrated into normative and ethical questions. That is why it is important to remain alert and develop an epistemological first aid to prevent the worst of such cases.

In closing let me use a different example that predates the fMRI and related imaging technologies. In 1937 Penfield and Boldrey published their first report on the so-called "homunculus." This first depiction, frequently reproduced in textbooks and on thousands of web pages, presumably gives the topography of the primary motor cortex. Based on the cortical stimulation of 126 (awake) patients, these collated reports suggested a comprehensive map. In 1950 however Penfield and Rasmussen published a second, presumably more accurate and updated version of their homunculus, also frequently reproduced in textbooks and on web pages (Penfield & Rasmussen, 1950). As Schott (1993) later noted, "It is unclear whether the authors appreciated the visual significance of the homunculi, but these figurines created a precedent which has had a major influence on subsequent forms of related graphic illustration" (p. 330). Despite the popularity of the homunculus diagrams and their omnipresence in textbooks, the representations were, as Schott argued, of limited scientific value (see also Schieber, 2001). Leaving aside the neuroanatomical details, it is now a much more complicated story, argues Catani (2017), "... the homunculus holds the key to the precise coding that results in the coordinated activation of peripheral muscles. But the colloquial use of the term bears the risk of mistakenly granting the homunculus an existence in the realm of neuroscience: this little man, like many other figures that may naively populate our collective imagination, is just a metaphor for the complex neurological mechanisms that we strive to comprehend in their entirety" (p. 3061). A representation that has helped explain the primary motor cortex to millions of undergraduates turns out to be a useful educational tool but not a scientific one. The tool has been used to explain somatotopic organization as shorthand for what we take the primary motor cortex to be. However, over an extended period of time the value of this figure has largely become a rhetorical device. Here is a kind of epistemological first aid in slow motion, the gradual dissipation of a favorite figure to generations of undergraduates and medical students.

It should be obvious that epistemological first aid is nothing less than a call to remain critical of all and sundry claims made by the neurosciences on behalf of psychological categories. As an ever-expanding technology it is helpful to remember that it relies as much on psychological categories as it does on the technologies it uses to code brain states.

## References

Anderson, M. L. (2014). *After phrenology: Neural reuse and the interactive brain*. Cambridge, MA: MIT Press.

Bell, D. (1974). *The coming of post-industrial society*. New York: Harper Colophon Books.

Bennett, M. R., Dennett, D., Hacker, P. M. S., Searle, J., & Robinson, D. (2007). *Neuroscience and philosophy: Brain, mind, and language*. New York: Columbia University Press.

Bennett, M. R., & Hacker, P. M. S. (2003). *Philosophical foundations of neuroscience*. New York: Wiley.

Catani, M. (2017). A little man of some importance. *Brain, 140*, 3055–3061.

Choudhury, S., & Slaby, J. (2012). *Critical neuroscience: A handbook of the social and cultural contexts of neuroscience*. Chichester: Blackwell.

Coey, C. A., Varlet, M., & Richardson, M. J. (2012). Coordination dynamics in a socially situated nervous system. *Frontiers in Human Neuroscience, 6*, 164. https://doi.org/10.3389/fnhum.2012.00164.

Darley, J. M., & Latané, B. (1968). Bystander intervention in emergencies: Diffusion of responsibility. *Journal of Personality and Social Psychology, 8*, 377–383.

De Rijcke, S., & Beaulieu, A. (2007). Essay review: Taking a good look at why scientific images don't speak for themselves. *Theory & Psychology, 17*, 733–742.

Dennett, D. C. (1988). The moral first aid manual. In M. S. McMurrin (Ed.), *Tanner lectures on human values* (Vol. VIII, pp. 120–147). Salt Lake City: University of Utah Press.

Dumit, J. (2004). *Picturing personhood: Brain scans and biomedical identity*. Princeton: Princeton University Press.

Dupré, J. (2001). *Human nature and the limits of science*. Oxford: Oxford University Press.

Fischer, P., Krueger, J. I., Greitemeyer, T., Vogrincic, C., Kastenmüller, A., Frey, D., et al. (2011). The bystander-effect: A meta-analytic review on bystander intervention in dangerous and non-dangerous emergencies. *Psychological Bulletin, 137,* 517–537.

Foster, J. A. (2013, July 1). Gut feelings: Bacteria and the brain. *Cerebrum*. Retrieved from http://www.dana.org/Cerebrum/Default.aspx?id=39496.

Furr, N., Nel, K., & Ramsøy, T. Z. (2019). Neuroscience is going to change how businesses understand their customers. *Harvard Business Review*. Retrieved from https://hbr.org/2019/02/neuroscience-is-going-to-change-how-businesses-understand-their-customers.

Gergen, K. J. (2010). The acculturated brain. *Theory & Psychology, 20,* 795–816.

Gigerenzer, G. (1991). From tools to theories: A heuristic of discovery in cognitive psychology. *Psychological Review, 98,* 254–267.

Greene, J., & Cohen, J. (2004). For the law, neuroscience changes nothing and everything. *Philosophical Transactions of the Royal Society of London B: Biological Sciences, 359,* 1775–1785.

Haidt, J. (2012, October 1). Jonathan Haidt on moral psychology. *Social Science Space*. Retrieved from https://www.socialsciencespace.com/2012/10/jonathan-haidt-on-moral-psychology/.

Hickman, L. A. (2001). *Philosophical tools for technological culture: Putting pragmatism to work*. Bloomington: Indiana University Press.

Hortensius, R., & de Gelder, B. (2014). The neural basis of the bystander effect—The influence of group size on neural activity when witnessing an emergency. *Neuroimage, 93*(Pt1), 53–58.

Hull, C. L. (1934). The concept of the habit-family hierarchy and maze learning: Part I. *Psychological Review, 41,* 33–54.

Jordan-Young, R., & Rumiati, R. I. (2012). Hardwired for sexism? Approaches to sex/gender in neuroscience. *Neuroethics, 5,* 305–315.

Latour, B. (1990). Technology is society made durable. *The Sociological Review, 38*(S1), 103–131.

Latour, B. (2004). How to talk about the body? The normative dimension of science studies. *Body & Society, 10,* 205–229.

Madroñal, N., et al. (2016). Rapid erasure of hippocampal memory following inhibition of dentate gyrus granule cells. *Nature Communications, 7.* Article no. 10923. Retrieved from https://www.nature.com/articles/ncomms10923.

Margolis, J. (1984). *Philosophy of psychology*. New York: Prentice-Hall.
McGrath, M. (2019, March 28). Good Samaritans after all. *Aeon*. Retrieved from https://aeon.co/essays/it-looks-like-human-beings-might-be-good-samaritans-after-all?utm_medium=feed&utm_source=rss-feed.
Panchin, A. Y., Tuzhikov, A., & Panchin, Y. V. (2014). Midichlorians—The biomeme hypothesis: Is there a microbial component to religious rituals? *Biology Direct, 9*. Retrieved from http://www.biologydirect.com/content/9/1/14.
Penfield, W., & Boldrev, E. (1937). Somatic motor and sensory representation in the cerebral cortex of man as studied by electrical stimulation. *Brain, 60*, 389–443.
Penfield, W., & Rasmussen, T. (1950). *The cerebral cortex of man*. New York: Macmillan.
Radder, H., & Meyen, G. (2012). Does the brain "initiate" freely willed processes? A philosophy of science critique of Libet-type experiments and their interpretation. *Theory & Psychology, 23*, 3–21.
Robinson, D. (1973). *The enlightened machine: An analytical introduction to neuropsychology*. New York: Dickinson.
Roiser, J. (2015). What has neuroscience ever done for us? *The Psychologist, 28*, 284–287.
Rose, N., & Abi-Rached, J. M. (2013). *Neuro: The new brain sciences and the management of the mind*. Princeton, NJ: Princeton University Press.
Rose, N., O'Malley, P., & Valverde, M. (2009, September 16). Governmentality. *Annual Review of Law and Social Science, 2*, 83–104. Sydney Law School Research Paper No. 09/94. Available at SSRN https://ssrn.com/abstract=1474131.
Schieber, M. H. (2001). Constraints on somatotopic organization in the primary motor cortex. *Journal of Neurophysiology, 86*, 2125–2143.
Schmitz, S., & Höppner, G. (2014). Neurofeminism and feminist neurosciences: A critical review of contemporary brain research. *Frontiers in Human Neuroscience, 8*. Article no. 546. https://doi.org/10.3389/fnhum.2014.00546.
Schott, G. D. (1993). Penfield's homunculus: A note on cerebral cartography. *Journal of Neurology, Neurosurgery and Psychiatry, 56*, 329–333.
Schwartz, S. J., Lilienfeld, S. O., Meca, A., & Sauvigné, K. C. (2016). The role of neuroscience within psychology: A call for inclusiveness over exclusiveness. *American Psychologist, 71*, 52–70.

Smith, R. (2010). *The will: Victorian and modern responses to the brain sciences*. Paper presented at the Symposium on Neuroscience and Human Nature, Wellcome Trust Centre for the History of Medicine at UCL. Retrieved from http://www.rogersmith.ru/papers.

Solymosi, T. (2014). Moral first aid for a neuroscientific age. In T. Solymosi & J. R. Shook (Eds.), *Neuroscience, neurophilosophy and pragmatism: Brains at work with the world* (pp. 291–317). New York: Palgrave Macmillan.

Stam, H. J. (2015). The neurosciences and the search for a unified psychology: The science and esthetics of a single framework. *Frontiers in Psychology, 6*. Article no. 1467. https://doi.org/10.3389/fpsyg.2015.01467.

Useem, J. (2017, July/August). Power causes brain damage: How leaders lose mental capacities—Most notably for reading other people—That were essential to their rise. *The Atlantic*. Retrieved from https://www.theatlantic.com/magazine/archive/2017/07/power-causes-brain-damage/528711/.

# Part II

## Applying Psychological Concepts to the Study of Science and Technology

# 5

# *"Groping for Trouts in a Peculiar River:"* Challenges in Exploration and Application for Ethnographic Study of Interdisciplinary Science

Lisa M. Osbeck and Nancy J. Nersessian

## Introduction: Interdisciplinary Science and Psychology

For the psychological researcher, interdisciplinary research labs constitute important sites for inquiry. They afford interpretation of the interplay of cognitive, social, cultural, and material dimensions of creative epistemic practices and constitute significant sites of learning. They are evolving communities, demonstrating the complex processes through which discourses, methods, artifacts, and representations from different disciplinary traditions interact (Nersessian, 2012, 2019a, in press; Nersessian, Newstetter, Kurz-Milcke, & Davies, 2002). The norms that structure their contexts and give rise to possibilities

L. M. Osbeck (✉)
Ann Arbor, MI, USA
e-mail: losbeck@westga.edu

N. J. Nersessian
Somerville, MA, USA





K. C. O'Doherty et al. (eds.), *Psychological Studies of Science and Technology*, Palgrave Studies in the Theory and History of Psychology,
https://doi.org/10.1007/978-3-030-25308-0_5

for producing knowledge are in negotiation, because scientists from different backgrounds and different normative traditions, enter the practice context armed with what are sometimes divergent ideas about what it means to do science well, and how to be a good scientist. Thus there are insights to be gleaned about how scientists navigate social rules and enact epistemic identities (Osbeck & Nersessian, 2017). There are many affordances, then—opportunities—for better understanding science, and with potential for valuable application to educational contexts.

In addition to opportunities, however, exploratory study of interdisciplinary laboratories presents substantial challenges to the researcher. Understanding *how* learning occurs requires a methodological focus on learning *process.* In turn, a focus on learning process calls for understanding the specific cognitive *practices* of learners as situated in—not abstracted from—the contexts of their learning (Brown, Collins, & Duguid, 1989; Greeno, 1998). Characterizing reasoning practices as situated within a particular disciplinary domain (e.g., a science) requires knowledge of that domain as well as its conventions of communication and notation. Acquiring this knowledge can be daunting for cognitive and learning science researchers who are equipped with tools to characterize practice, yet who lack familiarity with the discipline studied (e.g., the laboratory science). We provide in this chapter a description of our efforts to address two challenges that arose in the context of our investigation of four research labs in different fields of the bioengineering sciences, for the purposes of which we used methods adapted from the anthropological tradition of interpreting culturally situated meanings in natural habitats of human practice—the everyday life world. These challenges are reflected in our title. Drawn from Shakespeare's *Measure for Measure* (1623), what was intended therein as a salacious metaphor also captures poetically the situation of the ethnographer investigating the practices of a research laboratory in a new or developing science. The river is the laboratory of scientists whose practices we seek to understand. It is "peculiar" in two senses: first, in that practices targeted are in some aspects unique to that setting (peculiar as particular); and second, in that they are strange or unfamiliar to the investigator, even more so if the ethnographers are social scientists without extensive prior knowledge of the science investigated. "Trouts" are the insights we seek and which we pursue through systematic collection and analysis of field notes and interview data, and "groping" suggests that a good deal of exploration and trial and error is involved in the efforts, especially in the early stages, when the "foreign" nature of the research laboratory presents myriad complexities to understand. The implication of clandestine relations suggested by the metaphor in its literary context is also a nod to the situation of the psychologist or philosopher when she deliberately borrows methods from another discipline and its investigatory tradition (anthropology). Although borrowing methods from one discipline to use in another opens new avenues of questioning and investigation, it also risks the possibility that the discipline whose methods were borrowed might question the legitimacy of the borrowing. The investigation thus requires attending to the canons of good usage in the originating discipline, their justification, and the discourse around critiques, while reflectively adapting methods to the needs and standards of the borrowing discipline.

Yet adapting anthropological methods for the psychological investigation of laboratory science practice is not only justified but required by our conception of a working research laboratory as complex evolving cognitive-cultural system and by our research aim to investigate science practice as it naturally occurs in the everyday contexts of scientists' activity. Nancy Nersessian (philosopher and historian of science and cognitive scientist) and Wendy Newstetter (linguistic anthropologist), supported by the US National Science Foundation, designed and implemented the multiyear investigation of frontier, innovation-seeking science practices that involved a comparison of four different sites of interdisciplinary practice, though all in the bioengineering sciences. They proposed to describe the cognitive and learning practices specific to each laboratory and then to draw comparisons across them, enabling description and analysis of different configurations of interdisciplinary practice. The investigation was intended to provide insight into the nature of interdisciplinary research as well as the conditions that facilitate learning and productive collaboration in emerging interdisciplinary fields. An additional goal was to gain insights of this kind to develop and implement strategies for improving science education.

# Study of Laboratories

## Precedents and Adaptations

The challenge to understand the practices of an unfamiliar context always confronts the ethnographic team, and this is no less the case when the context in question is a newly developing interdisciplinary science. Where and how to even begin to understand cognitive practices in these settings requires both consultation of key methodological precedents and a creative adaptation of these precedents to fit the specific features of our own inquiry. We here briefly review a targeted sample of relevant precedents and describe our specific adaptations. We then describe two distinct challenges we faced and how we attempted to address them.

An under-recognized pioneer in the history of the qualitative study of science is Ian Mitroff, who analyzed interviews with 42 scientists involved in the lunar missions of the late 1960s. Mitroff called his published analysis of the interviews "a book about how science actually gets done" (1974, p. 2), and his approach "a social psychology of research" (p. 20). His interviews with NASA, university, and other government or private industry scientists probed the nature of means by which they framed and addressed their research problems as well as how they conceptualized science more broadly. Using the responses provided to critique the "storybook" image of science, Mitroff also offered insights into the role of emotion, disposition, "faith" and commitment, relationships with other researchers, cognitive style, and "issues that are related to scientists as persons" (1974, p. 43) in the practice of science, for which reason he considered his study to offer a description of scientists' subjectivity. However, his study focused on the personality of the individual scientist and did not analyze features of the material or social contexts of practice or consider how these contexts contribute to the personal dimensions of the research process.

By contrast, Latour and Woolgar's pioneering *Laboratory Life: The Construction of Scientific Fact* (2013/1979) focused on the "routine work carried out in one particular laboratory," and its method entailed "in situ monitoring of scientists' activity in one setting" (p. 27). The emphasis on the particularity of the setting directly counters the idea of "scientific method" as a set of standardized procedures indifferent to context, and also counters Mitroff's focus on the particularity of the researcher. Latour and Woolgar offered a detailed description of rhetorical strategies used in both formal and informal presentation of research results, including the formal (e.g., written summaries, presentations) and informal (e.g., lab meetings, conversations) use of numbers, graphs, and other representational formats to argue and convince, especially in relation to research controversies. We align with Latour and Woolgar in considering the social and material features of the context to be an important focus of analysis. We similarly regard the laboratory as a system of interaction, embedded within larger normative systems (the university, bioengineering, Western science) that provide the conditions for "sense-making" and rhetorical strategy. However, unlike Latour and Woolgar, we include the cognitive practices of the researcher, especially model-based reasoning, among the aspects of the system that must be understood (Nersessian, 2005).

In this vein, our research framework and strategy more closely resemble that of Hutchins (1995a, 1995b) and the broader tradition of *cognitive anthropology* it represents.[1] Hutchins investigated "cognitive practices" in "natural habitats," defined as "naturally occurring culturally constituted human activity" (1995b, p. xiii). His focus is on the "ecology of thinking, in which human cognition interacts with an environment rich in organizing resources" (p. xiv). Contrasting his approach with that of studying cognition in the artificial conditions imposed by controlled experimental studies of reasoning in laboratory settings, Hutchins depicts human problem-solving as "cognitive practice," implying that the uniquely human form of it is always moored by the cultural system that gives it meaning: brain, body, and environment are co-implicated (Hutchins, 2014). Hutchins' analysis focused on the specific in situ means by which complex problem-solving is accomplished by what he called a "socio-technical system": a group

[1]There are several threads of pioneering, mutually influential contributions that came together at that time (Engestrom, 1987; Lynch, 1985; Norman, 1988). Nersessian and Newstetter came to appropriate and develop the method from cognitive anthropologists Hutchins (1995a, b) and Lave (1988) and only later discovered the confluence of what had become by then separate endeavors.

(or individual) interacting with culturally produced artifacts and representations (graphs, charts, statistical formulae, written instructions, traditions of practice) solving problems toward the accomplishment of a predetermined goal (e.g., landing a plane, navigating a ship).

Our research group's investigation of bioengineering science laboratories is an adaptation of cognitive anthropology to the study of interdisciplinary science and its practices. Our multiyear investigation examined four sites of interdisciplinary practice: initially two in biomedical engineering (a tissue engineering lab and a neuro engineering lab), and later, two in systems biology (a biosystems computational modeling lab and a combined computational modeling and wet lab). We proceeded from the assumption that each lab is an evolving cognitive-cultural system that enables specific forms of complex problem-solving. We analyzed how specific features of the social and material environment are implicated in learning and reasoning practices. However, two important differences add complexity to the study of the laboratories in comparison with the context of the cockpit or ship. First, in contrast to Hutchins's studies, the laboratories we studied have ill-defined or only loosely defined problem-solving agendas. They are innovation-seeking communities with evolving goals, problem formulations, methods, and technologies. The graduate student researchers are learners whose evolving trajectories intersect with the other dimensions of what we call the lab-as-problem-space (Nersessian, Kurz-Milcke, Newstetter, & Davies, 2003). Second, although the focus of our analysis was, like that of Hutchins, on cognitive practices, in our analysis we adopted as a central unit of analysis "The acting person in normatively structured contexts of practice" (Osbeck & Nersessian, 2015, p. 29), which we view as an inherently integrated focus. Our interest, then, was not only in cognitive and social dimensions as these are traditionally understood but also in what we consider the "personal" dimension of practice, what we have called "the something else"—as implicated in but not reducible to either cognition or sociality, aligned with what Mitroff understood as subjectivity. This "something else" includes what is fundamentally a matter of "style," inflecting all one's activities with some mark of particularity that may or may not be part of the researcher's experience and self-representation. The challenge for science studies *and* psychology is to adequately integrate this dimension into accounts of practice without suggesting an isolated individualism. Acting persons are socially enculturated but uniquely constituted and storied beings, thus as a unit of analysis, "the acting person" encompasses all of these dimensions.

An additional assumption undergirding our study of laboratories is that to study cognition and learning specifically we are required to pay more attention to "transfer" than is typically thought to accompany ethnographic investigation. Research in the situated and distributed perspectives largely consists of observational case studies employing ethnographic methods. Their overarching objective, however, differs from sociocultural studies that aim mainly to ferret out the social, cultural, and material facets of a case. As cognitive science accounts, cognitive ethnographies need to move beyond richly nuanced details of the specific case toward a more general account of the regularities of cognition and how they function in human activity. Finally, because systems of practice (e.g., a laboratory) are nearly always embedded in larger cognitive and social systems, our aim was also to study and make claims with broader applicability to interdisciplinarity. That is, our concern was not only with particular features of each laboratory but about the epistemic situation of interdisciplinary science. We were intent on analyzing the epistemic features that characterize or distinguish interdisciplinary science from other situations in which learning and problem-solving take place, such as a more disciplinary bounded science. At the same time, we remained attentive to the organization and particular goals of each laboratory and considered how its particular social organization and particular problems and goals interface with the more general structures introduced by interdisciplinary practice. We were concerned both what is consistent across and distinct within the culture of the laboratories studied.

## Methods

We describe our methods in detail elsewhere (Nersessian, 2019b; Osbeck, Nersessian, Malone, & Newstetter, 2010, Chapter 2), and provide here only a brief overview:

Data Collection: In both phases of the study, we collected data of different kinds and from multiple sources: in situ field observations, interviews, laboratory tours, laboratory archival material, and field-relevant published literature. We took field notes while "hanging out" informally in the labs and conducted interviews mainly within the lab space. In the laboratory investigations, we have sought to understand (1) the reasoning and problem-solving strategies that drive the work of the research lab (cognitive and investigative practices) and (2) how lab newcomers apprentice to and learn these strategies (interactional practices).

**Data Collection.** For each lab, interviews were conducted with as many researchers as possible given time and resources. We were deliberate in interviewing participants with varying levels of education, differing amounts of research experience, and who represented each discipline engaged in collaboration. We used interviews with different formats (unstructured and semi-structured) and different foci. Some focused on researcher experience in the lab (interaction with others, significant learnings, personal and problem-solving goals, normative expectations, and progress); some collected biographical information; some recorded researcher accounts of collaboration with other researchers; others focused on devices and instrumentation. For a subset of participants in each lab we collected longitudinal data, conducting interviews at regular intervals on research project or learning progress. All interviews were transcribed, and field notes were organized chronologically for each lab. Note taking recorded impressions of social interactions and indications of formal and informal social hierarchies. We collected audio and visual tapes of formal lab meetings in which new findings and problems with ongoing research were presented and discussed. The ethnographic team also created a visual record of the spatial layout, consisting primarily of photographs and diagrams and collected participant white board and note paper sketches and doodles. Additionally, we collected and examined archival materials (grant proposals, slide presentations, paper drafts, dissertation proposals, conference posters), important in the lab as a whole and to particular researchers.

**Analysis.** During both phases of the study, we first conducted fine-grained open ("inductive") coding on a small set of interviews to develop our initial codes. Specifically, we selected a subset of interviews from each lab and analyzed progressively, line by line, from beginning to end, with the aim of providing an initial thematic (interpretive) description for all meaning units identified. We initially coded within labs, using different sets of ethnographers for each lab. This was followed by a phase of making comparisons across labs within each area. During this second phase, we refined existing codes and developed new ones as they emerged from the comparison. We also compared codes arising from transcripts with themes analyzed from field notes in order to amplify or refine emergent codes. We also analyzed the subset of interviews collected at regular intervals by constructing a narrative case study to describe the problem-solving or learning trajectory of the selected research scientists over time. Finally, we used cognitive-historical analysis, a method developed by Nersessian and colleagues to analyze the development of specific research problems, technologies, key concepts, models, and methods used in each laboratory (Nersessian, 1987, 1992, 2008).

**Rigor and Accountability**. We attended to possible charges that data collected in an ethnographic study might not be fully representative and that all interpretive analyses are subjective. We respond to this charge by pointing out that unlike customary practices where the ethnographer is a single researcher, we practiced what we dubbed "team ethnography." More than one ethnographer had responsibility for observations and interviews in a given lab, and the more senior members of our group worked across the labs. Our weekly research group meetings provided the venue for scrutinizing and evaluating interpretations interactively, both generating new ideas and reaching consensus on coding, theme development, and other forms of data interpretation. The meetings also provided the opportunity to collectively relate our findings to appropriate cognitive, sociocultural, and philosophical theoretical frameworks. Our research group varied in size and composition over time (undergraduates through senior faculty), but remained highly interdisciplinary and thus provided multiple lenses through which we could understand the data.

Codes that had emerged through the inductive process were tested for their applicability and conceptual fit with new interviews and from field note recordings. We also enlisted an independent auditor to evaluate the plausibility of coding concerning a subset of interviews.

## Important Findings

We generated numerous codes and a set of higher order codes that corresponded to forms of activity: Modeling, Framing, Positioning, Offering historical narrative (lab and personal); Expressing or enacting identity; Expressing emotion/affect/motivation/desire; Displaying experience of agency; Acknowledging norms, Managing Complexity. Among the most important findings was that researchers of all levels and across all labs gave accounts of their problem-solving that we interpreted as indicative of model-based cognitive practices. Conceptual models, physical in vitro models of in vivo phenomena, mathematical models, and computational models integrating biology and engineering abound in the investigative practices of bioengineering scientists. We characterized researchers as engaging "interlocking models" (Nersessian, 2009; Nersessian & Patton, 2009; Osbeck & Nersessian, 2006). At the level of activity, we distinguished *model-based understanding*, defined as referencing, demonstrating, explaining, giving evidence of comprehension in terms of an organized representation; *model-based reasoning*, defined as providing evidence of reasoning through model construction and manipulation; marked by inferences ("if-then," "thus," "it seems like," "maybe," "I suspect," "so," etc.); and *model-based explaining*, marked by the use of models to communicate a complex idea to the interviewer. Explaining also included metaphorical or analogical comparisons used to aid the *interviewer's* understanding.

There are many other coding categories we related to these central modeling categories. We have detailed our findings for the biomedical engineering study in an authored book (Osbeck et al., 2010); we report findings from both phases of the study in numerous journal publications (a sample includes: Chandrasekharan & Nersessian, 2015, 2017; MacLeod & Nersessian, 2013a, 2013b, 2016; Nersessian, 2009, 2012, 2019b; Osbeck & Nersessian, 2006, 2015, 2017). For each analysis, our emphasis was on the integration of capacities in scientific problem-solving, and modeling itself as an integrated activity (integrating cognition, affect, culture, sociality, and agency), that is, the activity of acting persons.

## Application to Education

In addition to providing a framework for understanding the learning and innovative problem-solving practices in the bioscience laboratories, an important goal of the ethnographic study was to generate insights that would be useful for application to science education, to inform instructional design in classrooms and instructional labs. Our goal has been to understand both the challenges to learning and what makes for successful learning in complex settings of STEM practice and use that understanding to design educational environments that support complex learning in formal instructional settings. That is, we sought to inform curriculum development for biomedical engineering, with the broader goal to position students to engage the kind of thinking scientists actually do. Unfortunately, for the ethnographer, the "groping for trouts" metaphor pertains to this phase of the study as well, especially when tasked with the goal of developing concrete measures for *evaluating* whether such thinking is taking place in the classroom. With the bioengineering faculty, we designed three types of introductory courses to promote integrative model-based reasoning and problem-solving during our investigations: small classroom (Newstetter, 2005), experimental lab (Newstetter, Behravesh, Nersessian, & Fasse, 2010), and computational modeling lab (Voit, Newstetter, & Kemp, 2012). Our "translational" approach of investigating authentic practices and translating them into learning experiences through design-based research should be applicable across STEM fields. In the next section, we briefly describe some of the challenges we faced in designing and evaluating the first course—an introductory biomedical engineering course informed by the ethnographic study—and innovative strategies we developed to assess "thinking like a biomedical engineer" by students enrolled in the new course.

### Model-Based Reasoning in Bioengineering Education

As noted, our ethnographic studies of practices in BME laboratories led us to characterize cognitive practices in these settings as largely model-based. Based on our findings, the primary educational goal of our project was to assist students in developing a versatile and informed understanding of models and enabling them to engage a model-based approach to problem-solving spontaneously. Therefore in partnership with faculty from biomedical engineering, our team designed a first course based on genuine research problems and practices that would demand and foster utilization of and engagement with qualitative and mathematical modeling. We reasoned that *problem-based learning* (PBL) environments would foster model-based reasoning. In PBL, students learn by working through realistic but often ill-structured problems requiring them to conduct self-directed searches, integrate information from multiple disciplines, and reflect on their own experiences (Capon & Kuhn, 2004; Hmelo-Silver, 2004).

To arrive at a stable but flexible version of this course, we used a *design-based* research approach in which we ran a variety of test problems generated by biomedical engineering faculty through several iterations. The course enrolled approximately 80 students in the fall semester, 180 in the spring, and 30 in the summer. At the start of the semester students were divided into problem-driven learning sections of 8 students who meet three hours a week with a faculty facilitator to report on group research they have completed outside of class, to apply and integrate that new information and, as a team, to forge solution strategies to ill-constrained, ill-structured problems. As we observed in the engineering research laboratories, a compelling research problem does not merely situate or anchor learning; rather it compels, provokes, and drives learning forward. A relentless need to make progress in a complex problem space is what we tried to replicate in the classroom.

Each problem assigned revealed a different facet of biomedical engineering, from screening and detection of cancer using technologies that span protein changes (proteomic strategies), to experimental design for detecting sources of error in biometric devices, to mathematical modeling and computational simulation of physical systems as a method of hypothesis testing. The problems were open-ended, ill-constrained, and ill-structured, demanding an investigation of the intersection between technology and human physiology. The problems required students to integrate biomedical-engineering dimensions into their problem-solving from the outset of their educational experiences.

In turn, each problem requires the students to discover the resources they will need to solve the problem, to move from the qualitative to the quantitative, and to develop analytical frameworks from the data. The course goal is to have the student teams practice engineering problem-solving with the facilitation of a faculty member.

## Assessment of Model-Based Learning in PBL Classrooms

Our—and the BME faculty's—informal observations supported the claim that the PBL classroom design facilitated model-based reasoning, but we wrestled with an important assessment issue: how to distinguish between students' understanding of models (what is a model, how is it used in research) from students' spontaneous *construction and use* of models in the context of problem-solving. We developed several iterations of a self-report questionnaire before concluding that any such questionnaire measured only students' understanding of what a model is and how it functions rather than informing us of the extent to which students engaged in the process of strategic model-based reasoning.

Our final assessment strategy was to present students with a novel complex problem closely linked to the kind of problem they had encountered during the course. They had more than two hours to address the problem in writing in the context of a formal examination, graded by their facilitator. We did not look at the grading but examined a random selection of student exam responses to evaluate the forms of reasoning demonstrated. An example of the problems we posed is the following.

<u>Problem statement:</u>
*Systemic lupus erythematosus (SLE), commonly known as lupus, is a disabling, autoimmune disease that can lead to significant morbidity and mortality, particularly from renal and cardiovascular disease. There is growing concern that environmental factors may play a role in the development of this disease, specifically, exposure to respirable crystalline silica (quartz), a common mineral found in rocks and soil. How would you, as a biomedical engineer, investigate the potential link between this disease and quartz?*

*Directions for answering the question:*
*This final exam is an opportunity for you to demonstrate what you have learned about BME approaches to complex problem solving over the semester. Your grade will depend on how well you are able to use what you practiced in the three-semester problems. A suggested plan of attack is:*

1. *Identify relevant strategies, frameworks, content or concepts from tackling the semester problems that would be relevant here. Make sure to address human subjects' implications.*
2. *Make a list of the resources from each problem in the semester that you plan to use in tackling this problem.*
3. *Using these resources, present a coherent problem-solving strategy that articulates your approach to this problem. Use headings, sub-headings bullets or other devices in the formatting for maximum clarity and readability.*

## Developing Codes

Three raters coded 100 final exams from the BME 1300 course using a thematic coding system developed to characterize the problem-solving strategy exhibited in the exam responses. Rater #1 has a Ph.D. in biomedical engineering and had served as a facilitator in the course, though not in the semesters from which the exams were selected; rater #2 has expertise in cognitive science, learning sciences, and philosophy of science; and rater #3 has a background in qualitative methods, psychology, and philosophy of science. Raters 2 and 3 did not participate in teaching the BME 1300 course, and none of the three had involvement in the development of the final exam question, the grading of the final exam, or the collection of the exams to be scored. An additional rater (#4) with expertise in learning science, ethnographic analysis, and linguistics, and who facilitated the BME 1300 course, designed the two final exams in collaboration with other course facilitators and graded her sections, served as a consultant in determining the appropriateness of codes. Rater #4 also blindly coded a subset of the exams using the coding rubric developed by the other three raters.

Raters 1, 2, and 3 independently read the same subset of twenty exams, guided by the question of how best to characterize differences in problem-solving strategies. Raters then met to review and compare observations, focusing on differences evident across responses. Differences considered important by all three raters became a basis for codes. Raters generated codes by formulating a succinct and thoroughly plausible description of a selected passage, evaluating its particular context within the exam response. Descriptions were derived through detailed discussion about the possible significance of text passages and any plausible alternative interpretation of the portion of text in question. We adopted a code only when all three raters were in full agreement about its fit and relevance to the given passage. We revisited codes throughout the process of reading additional exams, refining some with further distinctions, revising others, adding and eliminating as our thinking evolved. Below, we provide verbatim samples from exam responses to illustrate the codes we developed.

The initial coding rubric developed is as follows:

**H = HYPOTHESIS**: Explicit reference to a hypothesis *or* an assertion of an expected (or conjectured) specific relation between two variables, expressed in the form of a proposition.

Examples:
*"After all of the preliminary research was obtained, a hypothesis, along with a null and potentially alternative hypothesis would be one that either a) followed an already existing idea on the determining link between quartz and SLE, or b) created a novel idea on the correlation between quartz and SLE"*

*"The next thing to do is to make a hypothesis on whether or not there is a potential link between lupus and quartz"*

**M = MODEL**: Statement expressing construction of an organized representation of a *range of possible relations* between variables, including structural or functional properties, mechanisms, and interactional patterns. This code also indicates statements that make direct reference to models, and instances of or references to charts or diagrams of relations, mathematical expressions, or a decision matrix.

Examples:
*"After analyzing all of my data and detecting possible patterns, I would begin in the creation of a scientific model with quartz, or the specific harmful elements of quartz as my inputs...."*

*"Use knowledge gained from research to devise a mathematical model. The model will have inputs and outputs based on information gathered. After probable model is completed, list assumptions and limitations of the model. Create a visual aid of the model to identify insight gained......."*

*"Once lupus and quartz are completely understood [i.e. from available background research], the next step would be to focus on the problem-specific details, as well as any pre-existing connections. Make note of any similarity between biological effects of quartz (if any) and the symptoms of lupus."*

*"I would first try to understand everything about SLE, from its molecular interactions with both foreign and local molecules to the larger picture of how it can lead to morbidity and even mortality...It would be important to know chemical properties... it would be crucial to know how quartz enters the body..."*

**ML = MODEL "LITE"**: Statement referencing an organized representation of relations but without specification of relational structure.

Examples:
*"Before proposing any experiments we would have to link lupus with quartz"*

*"Construct time-line for current patients – date, period of exposure, bodily effects"*

**T = TEST**: Statement specifying how a relation between variables will be tested

T-O: Test by Observational Study
T-E: Test by Experiment
T-L: Test by literature review

Examples:
*"Using the model I would create an experiment to test my model and its assumptions, inputs, outputs, and overall results."*

*"Design a plan of attack to carry out the experiment based on the hypothesis."*

Order of Codes: For each exam, raters used the code categories given above to outline the order of steps evident in the problem-solving strategy recorded by the student.

Thus, for example, a code of H > T-O > M reflects a problem-solving strategy in which a hypothesis is proposed, tested by observational methods (population studies), and then a model is constructed to explain the relations observed.

## Application of Codes to Evaluate Model-Based Learning

Using the coding rubric, the three raters then independently coded another set of twenty fall semester exams. Raters compared codes and brought questions and problems to be addressed by the coding group, refining codes as needed to account for problems and discrepancies encountered in the first round of coding, then providing for each code and subcode. The three coders then met with rater 4 to check the face validity of the codes and the coding strategy.

Raters 1 and 3 then independently coded the rest of the 50 fall exams and met to compare codes. The coding rubric was further refined and applied to 50 exams from the spring semester BME 1300 course.[2] Thus in total 100 exams were coded. Inter-rater reliability checks were

[2]Although it did not affect the scoring results we report here, the raters made additions to or further refinements of the codes based on the second round of coding include the following:

BR = Background Research (evidence of fact gathering without interest in relational structure of facts)

T-S = Test by Survey (self-report)

T-M = Test by mathematical model

T-U = Test Unspecified = reference to a test without design details

T-E-U = Test-Experiment-Unspecified = reference to experiment without details on subjects or manipulations.

H-R = Hypothesis Referenced = Reference to having made a hypothesis but the hypothesis is never stated.

H-U = Hypothesis Unspecified = "Forming hypothesis" is clearly mentioned as a component of research design but details of hypothesis are not provided.

obtained by assessing the extent to which raters agreed on the content and order of codes in relation to exam responses. The rate of agreement between raters was .9 for the second set of exams. Inter-rater reliability data for the first set of exams was not analyzed given that the codes evolved in accordance with group discussions. However, raters 2 and 4 independently sampled 20 exams from the set of 100 and compared coding with that of raters 1 and 3. In each case a high degree of correspondence between raters was evident.

Raters 1 and 3 then applied the refined coding rubric to an additional set of exams with a different problem:

Problem statement:
*In the last ten years, there has been an increased focus on how physical activity might influence the cognitive vitality of older adults. Some of these studies have examined changes in cognition within the normal range, whereas others have asked whether lifestyle factors such as physical activity reduces the risk or delays the onset of age-related diseases, such as Alzheimer's or vascular dementia. How would you as a biomedical engineer, investigate a potential link between Alzheimer's and physical activity?*

Examples of coded text passages in relation to Hypothesis and Model codes are as follows:

**H**:
*"If rats perform more physical activity, then their chances of getting an age-related disease will decrease"*

*"I would make direct assumptions here that the level of cognitive function correlates directly to the risk for age-associated disorders".*

**M**:
*"look at how the brain works cognitively; that is, the inner workings such as inter-neural connections, brain wave patterns; then look at how/why physical activity could affect these patterns.*

## Student Exam Results

The table below summarizes results of our coding analysis, organized by percentage of exam responses reflecting various problem-solving strategies:

| | Fall exams | Spring exams | Overall |
|---|---|---|---|
| M (model- based) Note: Includes all cases of starting inquiry with a model-like strategy, including M-L > H and cases in which a model is not followed by a hypothesis, such a M > T | 41/50 = 82% | 45/50 = 90% | 86/100 = 86% |
| M > H structure (includes Model-Lite) Includes all cases of model followed by a hypothesis (or call for a hypothesis) | 29/50 = 58% | 33/50 = 66% | 62/100 = 62% |
| M > H structure, Model-Lite removed | 22/50 = 44% | 25 = 50% | 47/100 = 47% |
| BR > H (BR-background research) | 4/50 = 8% | 5/50 =10% | 9/100 = 9% |
| H | | 3/50 = 6% | 3/100 = 3% |
| TE > H (TE-test experiment) | | 1/50 = 2% | 1/100 =1% |

Our coding strongly suggests that a majority of students (88%) used a model-based approach in their response to the open-ended final exam question. When the less developed examples of a model-based approach are removed (the "model-lite" instances), the effect is more modest. However, model-lite use demonstrates that students had learned that one does not move straight to making a hypothesis without first considering some structural, functional, or behavioral relationships among the potential variables. Thus, our results are vastly different from the percentages one would expect from a group of students exposed only to the received view of scientific problem-solving, for which we would expect a specific hypothesis at the beginning of the problem-solving strategy.

Exam responses reflected a model-based approach in one or more of the following ways: First, students adopted a problem-solving strategy whereby preliminary reflection and organized representation of structural, behavioral, or functional interrelations of the variables preceded the formulation of specific hypotheses. Second, they noted the need for an organized understanding of observed interrelations among variables and possible mechanisms responsible for these observations. Third, problem-solving strategies reflected an initial interest in understanding the variables deeply and fundamentally: their structure, functions, and interactions.

Although not all students demonstrated a model-based approach of this kind, the high percentage of those who did supports our assertion that the BME students were beginning to "think like biomedical engineers." The significance of this finding is best appreciated against what we would expect from students employing the received view of scientific problem-solving, characterized by the application of a hypothesis-driven scientific method. Our study design did not enable us to use a genuine control group for comparison. It was also not possible to use a pre-post research design here because it would make no sense to present a complex problem to incoming students and ask them to solve it like a biomedical engineer. However, it is fair to assume that most students would enter BME 1300 having been exposed to (and possibly having practiced, e.g., for science fair) the received view. What we characterize as the received view of scientific problem-solving stems from a long tradition in the philosophy of science that has influenced the textbook representation of scientific method. The Scientific Method as it appears in science textbooks, online resources, and research methods courses is a process by which one begins by narrowing general ideas about a topic of interest into a hypothesis through some unspecified process. The hypothesis is then subjected to at least one test with the use of a specific data set. Thus, we can compare problem-solving strategies observed in the two BME1300 exams to the strategy implicated by the received view, which would be hypothesis rather than model driven, and also the typical hypothetico-deductive PBL strategy:

- The Scientific Method as in traditional texts:
  H > T (possibly, > M) or BR > H > T
- Typical PBL reasoning cycle:
  BR > H > T (> M)
- Primary reasoning pattern expected of our PDL students:
  M > H > T (> M)

## Summary and Conclusion

The study of scientific reasoning in real-world contexts affords insight into the complex multidimensional processes by which research takes place. These insights are valuable not only for their own sake; they have the potential to enhance educational practice, including STEM classroom design. Yet there are also challenges at every stage. Honest confrontation of the challenges leads us to both acknowledge and embrace the innovative core of ethnographic inquiry, even as we strive to follow systematic procedures within it. In this chapter, we described two specific sets of challenges and how we addressed them. The first concerned the initial problem of interpreting the practices of an emerging interdisciplinary science, especially when no previous studies provide models precisely suited to our investigative goals. The second challenge was evaluating the extent to which applications made from the ethnographic study were implemented adequately in classroom design. In the context of describing our response to these challenges, we provided details of our process for developing codes to characterize laboratory practices and our development of a strategy for evaluating model-based reasoning in student exams. These efforts underscore the importance of the ethnographic emphasis on "researcher as instrument," even in the effort to provide a rigorous and replicable basis for analysis and application.

The ethnographic team who conducted the study included representatives from philosophy of science, cognitive science, linguistic anthropology, computer science, psychology, public policy, bioengineering, industrial design, women's studies, and psychoanalysis. Thus like the labs investigated, our ethnographic research team was itself an evolving cognitive-cultural system that adapted continuously to the differing interests and perspectives represented. Our working assumption was that though situated in the specific social and epistemic features of each laboratory, our analysis of cognitive practices was also informative *our* cognitive practices, especially because we are also "acting persons." That is, we assumed that it would inform psychological dimensions of interdisciplinary inquiry more broadly.

**Acknowledgements** We gratefully acknowledge the support of the National Science Foundation REC0106773, DRL0411825, and DRL097394084 in conducting this research. We appreciate the contributions of all the members of the Cognition and Learning in Interdisciplinary Research Cultures research group to the project and, in particular, Wendy Newstetter (co-PI) and Kyla Ross for their contributions to the assessment of MRB in PBL as detailed in this paper. We thank the directors and members of our research labs for welcoming us into their labs and allowing us numerous interviews. An earlier version of this paper was presented for the APA Division 5 mini-conference, with support provided by the Educational Testing Service (ETS) and Multi-Health Systems (MHS), Princeton, NJ, March 2018.

## References

Brown, J. S., Collins, A., & Duguid, P. (1989). Situated cognition and the culture of learning. *Educational Researcher, 18*(1), 32–42.

Capon, N., & Kuhn, D. (2004). What's so good about problem-based learning? *Cognition and Instruction, 22*(1), 61–79.

Chandrasekharan, S., & Nersessian, N. J. (2015). Building cognition: The construction of external representations for discovery. *Cognitive Science, 39,* 1727–1763.

Chandrasekharan, S., & Nersessian, N. J. (2017). Rethinking correspondence: How the process of constructing models leads to discoveries and transfer in the bioengineering sciences. *Synthese*, 1–30.

Engestrom, Y. (1987). *Learning by expanding: An activity-theoretical approach to developmental research*. Helsinki, Finland: Orienta-Konsultit.

Greeno, J. G. (1998). The situativity of knowing, learning, and research. *American Psychologist, 53*(1), 5–26.

Hmelo-Silver, C. E. (2004). Problem-based learning: What and how do students learn? *Educational Psychology Review, 16*(2), 235–266.

Hutchins, E. (1995a). How a cockpit remembers its speeds. *Cognitive Science, 19*(3), 265–268.

Hutchins, E. (1995b). *Cognition in the wild*. Cambridge, MA: MIT Press.

Hutchins, E. (2014). The cultural ecosystem of human cognition. *Philosophical Psychology, 27*(1), 34–49.

Latour, B., & Woolgar, S. (2013). *Laboratory life: The construction of scientific facts*. Princeton, NJ: Princeton University Press. Originally published 1979.

Lave, J. (1988). *Cognition in practice: Mind, mathematics, and culture in everyday life*. New York: Cambridge University Press.
Lynch, M. (1985). *Art and artifact in laboratory science: A study of shop work and shop talk in a research laboratory*. London: Routledge and Kegan Paul.
MacLeod, M., & Nersessian, N. J. (2013a). Building simulations from the ground-up: Modeling and theory in systems biology. *Philosophy of Science, 80*, 533–556.
MacLeod, M., & Nersessian, N. J. (2013b, in press). Coupling simulation and experiment: The bimodal strategy in integrative systems biology. *Studies in History and Philosophy of Biological and Biomedical Sciences, 44*(4), 572–584.
MacLeod, M., & Nersessian, N. J. (2016). Interdisciplinary problem-solving: Emerging modes in integrative systems biology. *European Journal for Philosophy of Science, 6*(3), 401–418.
Mitroff, I. I. (1974). *The subjective side of science: A philosophical enquiry into the psychology of the Apollo moon scientists*. Amsterdam and New York: Elsevier.
Nersessian, N. J. (1987). A cognitive-historical approach to meaning in scientific theories. In *The process of science* (pp. 161–179). Dordrecht, The Netherlands: Kluwer Academic.
Nersessian, N. J. (1992). How do scientists think? Capturing the dynamics of conceptual change in science. In R. Giere (Ed.), *Minnesota studies in the philosophy of science* (pp. 3–45). Minneapolis: University of Minnesota Press.
Nersessian, N. J. (2005). Interpreting scientific and engineering practices: Integrating the cognitive, social, and cultural dimensions. In M. Gorman, R. D. Tweney, D. Gooding, & A. Kincannon (Eds.), *Scientific and technological thinking* (pp. 17–56). Hillsdale, NJ: Lawrence Erlbaum.
Nersessian, N. J. (2008). *Creating scientific concepts*. Cambridge, MA: MIT Press.
Nersessian, N. J. (2009). How do engineering scientists think? Model-based simulation in biomedical engineering research laboratories. *Topics in Cognitive Science, 1*(4), 730–757.
Nersessian, N. J. (2012). Engineering concepts: The interplay between concept formation and modeling practices in bioengineering sciences. *Mind, Culture, and Activity, 19*(3), 222–239.
Nersessian, N. J. (2019a, in press). Creating cognitive-cultural scaffolding in interdisciplinary research laboratories. In A. C. Love & W. C. Wimsatt (Eds.), *Beyond the meme: Development and structure in cultural evolution*. Minneaoplis: University of Minnesota Press.
Nersessian, N. J. (2019b, forthcoming). Interdisciplinarities in action: Cognitive ethnography of bioengineering sciences research labs. *Perspectives on Science*.

Nersessian, N. J., Kurz-Milcke, E., Newstetter, W., & Davies, J. (2003). Research laboratories as evolving distributed cognitive systems. In D. Alterman & D. Kirsch (Eds.), *Proceedings of the Cognitive Science Society 25* (pp. 857–862). Hillsdale, NJ: Lawrence Erlbaum Associates.

Nersessian, N. J., Newstetter, W., Kurz-Milcke, E., & Davies, J. (2002). A mixed-method approach to studying distributed cognition in evolving environments. In *Proceedings of the International Conference on Learning Sciences* (pp. 307–314). Hillsdale, NJ: Lawrence Erlbaum Associates.

Nersessian, N. J., & Patton, C. (2009). Model-based reasoning in interdisciplinary engineering: Two case studies from biomedical engineering research laboratories. In A. Meijers (Ed.), *Philosophy of technology and engineering sciences* (pp. 678–718). Amsterdam: Elsevier Science.

Newstetter, W. C. (2005). Designing cognitive apprenticeships for biomedical engineering. *Journal of Engineering Education, 94*(2), 207–213.

Newstetter, W., Behravesh, E., Nersessian, N. J., & Fasse, B. (2010). Design principles for problem driven laboratories in biomedical engineering education. *Annals of Biomedical Engineering, 38,* 3257–3267.

Norman, D. A. (1988). *The psychology of everyday things*. New York: Basic Books.

Osbeck, L. M., & Nersessian, N. J. (2006). The distribution of representation. *Journal for the Theory of Social Behaviour, 36*(2), 141–160.

Osbeck, L. M., & Nersessian, N. J. (2015). Prolegomena to an empirical philosophy of science. In *Empirical philosophy of science* (pp. 13–35). Cham: Springer.

Osbeck, L. M., & Nersessian, N. J. (2017). Epistemic identities in interdisciplinary science. *Perspectives on Science, 25*(2), 226–260.

Osbeck, L. M., Nersessian, N. J., Malone, K. R., & Newstetter, W. C. (2010). *Science as psychology: Sense-making and identity in science practice*. New York: Cambridge University Press.

Shakespeare, W. (1623). *Measure for measure*. London: Folio. http://shakespeare.mit.edu/measure/full.html.

Voit, E. O., Newstetter, W. C., & Kemp, M. L. (2012). A feel for systems. *Molecular Systems Biology, 8*(1), 609.

# 6

# Scientists as (Not) Knowing Subjects: Unpacking Standpoint Theory and Epistemological Ignorance from a Psychological Perspective

Nora Ruck, Alexandra Rutherford, Markus Brunner and Katharina Hametner

## Positioning Psychology in Relation to Feminist Science and Technology Studies

Within the broad and diverse field of Science and Technology Studies (STS), feminist STS applies the insights of feminist theory and epistemology to the study of science and technology (see Creager, Lunbeck, & Schiebinger, 2001; Mayberry, Subramaniam, & Weasel, 2001).

N. Ruck (✉) · M. Brunner · K. Hametner
Sigmund Freud Private University Vienna, Vienna, Austria
e-mail: nora.ruck@sfu.ac.at

M. Brunner
e-mail: brunner@agpolpsy.de

K. Hametner
e-mail: Katharina.Hametner@sfu.ac.at

A. Rutherford
Department of Psychology, York University, Toronto, ON, Canada
e-mail: alexr@yorku.ca



K. C. O'Doherty et al. (eds.), *Psychological Studies of Science and Technology*, Palgrave Studies in the Theory and History of Psychology,
https://doi.org/10.1007/978-3-030-25308-0_6

Work within feminist STS has elucidated the relationships between gender and science, arguing not only that science is a social activity imbued with gender dynamics (inequities, sexism, androcentrism, cultures of masculinity, etc.), but that scientific models, theories, and knowledge are deeply imbued with gender (Haraway, 1989; Keller, 1985; Martin, 1991; Merchant, 1980; Schiebinger, 1993). Feminist studies of technology have demonstrated how the association of technology with masculinity has discouraged women from entering fields such as computer science and engineering. More recently feminist technoscience scholars have focused on the mutual shaping of gender and technology, regarding neither technology nor gender as immutable but rather co-constitutive (see Wajcman, 2007).

The relation between psychology and feminist STS has so far primarily been restricted to critique, that is, feminist STS scholars have subjected psychology to feminist and gender analysis. For example, Haraway's classic work *Primate Visions* explored the gendered and heteronormative dynamics and scientific/engineering vision of psychologist Robert Yerkes's Yale Laboratories of Primate Biology. Haraway also examined the monkey "mother-love" research of Harry Harlow at the University of Wisconsin-Madison to demonstrate how Harlow could "design and build experimental apparatus and model the bodies and minds of monkeys to tell the major stories of his culture and his historical moment" (Haraway, 1989, p. 231). Primatology and sociobiology, interdisciplinary fields that include psychology, have been fruitful objects of study for feminist STS scholars (see also Hrdy, 1981, 1999; Ruck, 2016).

There is also a small but growing body of work by historians of psychology that demonstrates how psychological theories, research designs, and the boundaries of scientific psychology have drawn on, maintained, and perpetuated gender stereotypes (e.g., Hegarty, 2013; Morawski, 1985; Nicholson, 2001, 2011; Rutherford, 2015; Shields, 2007). There are furthermore examples of theories/concepts from STS being applied to understand how women's experiences, such as date/acquaintance rape (Rutherford, 2017), post-partum depression (Held & Rutherford, 2012), and menstrual synchrony (Pettit & Vigor, 2015), have been realized and circulate through scientific and popular discourse.

However, psychology has rarely been used to advance empirical and theoretical research in STS despite the fact that a few of the earliest contributions to feminist science studies drew on psychological theories and concepts to understand how the dominant objectivist conceptualization of science is linked to masculinity. Notably, Evelyn Fox Keller—referencing the work of Nancy Chodorow, Dorothy Dinnerstein, and Jessica Benjamin—used feminist object relations theory to posit that intrapsychic developmental processes combined with cultural influences led to the association of masculinity with separation, autonomy, objectivity, and domination, while femininity became associated with subjectivity and interdependence. Inasmuch as the goals of positivist science include objectivity (which involves separation of subject from object) and the domination of nature, these must be regarded as a projection of masculinity onto science. As she wrote in the early 1980s, "I suggest that the impulse towards domination does find expression in the goals (and even in the theories and practices) of modern science, and argue that where it finds such expression the impulse needs to be acknowledged as projection" (Keller, 1982, p. 598; see also Keller, 1978). Despite Keller's work, the cross-fertilization between psychological theory and feminist STS has not been nearly as extensive as that between feminist STS and philosophy, history, sociology, and anthropology. As mentioned in the introduction to this volume (insert reference when it becomes available), this may be because of psychology's overreliance on positivist approaches to knowledge production in contrast to the historical, critical, constructionist, qualitative, and theoretical approaches embraced by the other disciplines upon which feminist STS draws.

In tandem with feminist critiques of science that unfolded over the 1970s and 1980s, psychologists themselves developed approaches to science that drew on developments in the sociology of knowledge (e.g., Sherif, 1979; Unger, 1983). This strand of critique resulted in a branch of feminist psychology that draws on constructionist, critical, historical, and qualitative approaches. This branch has had uneven uptake internationally, with strongholds in Canada, the UK, Australia, and New Zealand, but with comparatively less uptake in the United States where feminist empiricism and the focus on sex/gender differences has driven psychologists' research agendas (for an overview of international developments, see Rutherford, Capdevila, Undurti, & Palmary, 2011). Within this feminist empiricist framework, there has been a small body of work that examines how cognitive processes such as implicit and ingroup bias interact with scientists' own gender in influencing research results and interpretations (e.g., Eagly, 2012). And of course there is a large body of psychological research exploring the factors related to women's continued underrepresentation in STEM fields. However, as Kumar (2012) notes, while psychological perspectives have much to offer the study of gender-science and gender-technology relationships, psychology is largely absent from STS.

It is within this context that we place our present chapter. Here, we take up the question not of how individual cognitive processes are brought to bear on gender theory or scientific research on gender, but rather of how several existing concepts that are central to feminist STS—such as standpoint theory/epistemology, epistemologies of ignorance, reflexivity, and intersectionality—might be further developed if psychological processes were considered and applied to understanding the subjectivities of scientists. For example, in the case of standpoint theory/epistemology, how does a scientist's social location translate into knowledge or lack thereof? As Harding has repeatedly emphasized, standpoint theory encourages consideration of the relationship between experience and knowledge as mediated through one's set of social locations (Harding, 1993). A standpoint is not a given, it is acquired through struggle. Conversely, the absence of knowledge among the dominant group that standpoint theory seeks to challenge also arises from specific social locations and experiences and is an active, rather than a passive process. How does this happen? And what does this mean for knowledge production in science, and by scientists?

## Theories of Not Knowing: Feminist Standpoint Theories and Epistemological Ignorance

Feminist standpoint theories and, subsequently, critical race and feminist reflections on epistemological ignorance, offer psychologically relevant insights into blind spots in knowledge and knowledge production. Feminist standpoint theories emerged not least from feminist consciousness-raising groups of the late 1960s and 1970s in which women (re) claimed epistemic authority over their lives (Mendel, 2009). Noting that standpoint theories also evolved in ethnicity-based, queer, and anti-imperial social justice movements, feminist philosopher Sandra Harding describes standpoint theory as an "organic epistemology, methodology, philosophy of science, and social theory that can arise whenever oppressed people gain public voice" (Harding, 2004, p. 3).

According to Ian Parker, standpoint theory "reverses and transforms" what a "crass conspiratioral form of Marxism" (p. 722) calls *false consciousness*, i.e., the adopting by the working class of the ruling class' ideological understanding of the world. Standpoint theories insist, however, that false consciousness applies at least as much or *even more* to the ones in power as it does to the dispossessed, for a standpoint not only discloses to us but also conceals the world from us. Indeed, standpoint theories allow for the argument that the position of those in power provides only a limited and distorted perspective of social reality while the standpoint of subjugated groups offers a more complete and less distorted view. Harding calls this the exercise of "strong objectivity" (Harding, 1993), and stipulates that it is actually *strong reflexivity* that is required in order to maximize this kind of objectivity: "Strong objectivity requires that the subject of knowledge be placed on the same critical, causal plane as the objects of knowledge. Thus, strong objectivity requires what we can think of as 'strong reflexivity'" (p. 69). That is, knowledge producers (i.e., scientists, too) must actively work to become aware of how their social location impedes their ability to perceive aspects of social reality that are accessible to others.

A standpoint is related to the social position of the knower (e.g., the social position as "woman" or as "scientist") but it is not identical with it (see also Rutherford, Sheese, & Ruck, 2015). It mediates between social position and knowledge not least by way of experience. Gender, for example, is a relevant analytical category for standpoint theorists because in most societies, the lives and daily practices of individuals are arranged differently along gender lines, affording women with experiences and possibilities that differ from men's. The participation of women of color in both dominant and marginalized cultures offers the epistemological advantage of recognizing a multiplicity of standpoints while remaining cognizant of power relations invisible to the dominant group. Patricia Hill Collins calls this double experience the "outsider within" status of Black women (1986, 14) while Aída Hurtado speaks of a "mestiza (hybrid) consciousness" (2010, 33) that affords Latina women and other women of multiple social worlds with "multiple lenses" (ibid., title).

The concrete advantages and disadvantages of a given epistemic situation or location depend on the kind of knowledge pursued or on the particular epistemic objective in question (Alcoff, 2007). It is thus vital to emphasize that feminist standpoint arguments pertain mostly to knowledge about systems and consequences of oppression the knower is part of. A standpoint is then always a political project and as such it is a goal rather than a given. Oppressed groups begin to struggle for a standpoint when they learn to "turn an oppressive feature of the group's conditions into a source of critical insight about how the dominant society thinks and is structured" (Harding, 2004, p. 7).

This aspect of standpoint theory can be used to explain why in most countries, scientific theories about gender inequities are so strongly anchored in women's movements: In most scientific disciplines, including psychology, women scientists' concrete experiences with oppression and/or discrimination were vital for turning scientists' attention to these very oppressive social mechanisms (see for the relation between feminist psychologies, gender inequities, and feminist activism in different countries Rutherford et al., 2011). In other words, standpoint theories can help to both theorize and advance the scientific contributions by marginalized and/or oppressed social groups.

"Epistemological ignorance," a concept rooted in critical race theory, by contrast, can be drawn upon to understand how and why dominant groups often demonstrate systematic blind spots in their scientific knowledge production. One such blind spot by White American, male historians of psychology was pointed out by pioneer historians of women in psychology Elizabeth Scarborough and Laurel Furumoto (e.g., 1987) when they observed that historians of psychology had erased the contributions of early women psychologists to the discipline from their historiographic accounts. While standpoint theory can help explain why it took women or, more specifically, feminist psychologists to uncover this erasure and to "re-place" women in the history of psychology (see Bohan, 1995), epistemological ignorance may elucidate the mechanisms of the historiographic erasure itself. A different example here drawn directly from scientific psychology (as opposed to its historiography) is late nineteenth- and early twentieth-century White male Eurocentric psychologists' theories of racial difference (and, for that matter, gender difference). Most of these scientists were unable—because of their privileged position—to see how their theories of racial science simply reflected the power relations already established in society. In a similar vein, writers such as Francis Galton (e.g., 1869) were unable to see how class structures could affect one's ability to achieve eminence. Galton thus concluded that the capacity for eminence was almost exclusively inherited, despite the fact that his eminent families all shared the same environment (Galton, 1869). This stands as a remarkable oversight on his part unless one appreciates how class was invisible to him given his own vaunted class position.

Reflections on "epistemological ignorance" have tried to understand the very ways and processes of not knowing more systematically (e.g., Alcoff, 2007; Mills, 2007; Tuana, 2006, 2008). Charles Mills launched these discussions in the late 1990s when he suggested that White supremacy is based on an epistemological contract which consists of *epistemological ignorance* (Mills, 1997). Shannon Sullivan and Nancy Tuana took Mills' work up from a feminist perspective. Differentiating various ways of not knowing, Tuana suggested the following taxonomy of ignorance: (1) knowing that we do not know, but not caring to know, (2) we do not even know that we do not know, (3) they do not want us to know, (4) willful ignorance, (5) ignorance produced by the construction of epistemically disadvantaged identities, and (6) loving ignorance (Tuana, 2006). As this taxonomy indicates, ignorance is a multifaceted phenomenon that calls for an epistemology in its own right (Alcoff, 2007), or what some have referred to as an "agnotology" (Proctor & Schiebinger, 2008). What is more, Tuana's taxonomy alludes to the fact that epistemological ignorance relates to systems of privilege and oppression as well as to psychological dimensions like cognition, affect, and the unconscious in complex ways.

Beyond classifying various expressions of ignorance, epistemologists have analyzed ignorance as not (only) a lack of insight or knowledge but as an epistemic practice in its own right (Alcoff, 2007).

Mills (1997, p. 18) called epistemological ignorance an "inverted epistemology," a "pattern of localized and global cognitive dysfunctions (which are psychologically and socially functional)," and a "group-based cognitive handicap" (2007, p. 15), which produces "the ironic outcome that whites will in general be unable to understand the world they themselves have made" (p. 18). Furthermore, he argues that the notion of epistemological ignorance is a "straightforward corollary of standpoint theory" for "if one group is privileged, after all, it must be by comparison with another group that is handicapped" (p. 15). While standpoint theories hold that those in positions of power have *less* interest in scrutinizing their own dominance critically, i.e., in seeing systems of dominance and oppression correctly, this change in perspective posits that they have indeed a "*positive* interest in 'seeing the world wrongly'" (Alcoff, 2007, p. 47).

## Psychologies of Not Knowing or Ignorance

In line with theories of epistemological ignorance, Parker has emphasized that for those in positions of power, "their partial view of the world corresponds with their own interests and obscures the operations of the very power they benefit from" (Parker, 2015, p. 724). This kind of epistemological ignorance has a *social function* in that it keeps systems of oppression in place (Mills, 1997, 2007). However, Mills alludes to the fact that epistemological ignorance is also *psychologically* functional.

We now exemplify how the epistemological obscuring of power mechanisms may work psychologically by turning to Gabriele Rosenthal's studies on the silence of Nazi perpetrators and their families, which suggest that, in the case of perpetrators or maybe of those in power more generally, ignorance functions as a psychological defense mechanism particularly against feelings of guilt. In interviews conducted with former Nazi perpetrators and their spouses as well as their children and grandchildren during the early 1990s, Rosenthal and her colleagues found family dialogues that veiled and denied the crimes of the parents and, sometimes, the entire parent generation (e.g., Rosenthal, 1998). Attending to the concrete psychological defense mechanisms in place, Rosenthal (1998) reconstructed three strategies of deflecting responsibility that were present in all three generations of perpetrator families.

A first strategy of deflecting responsibility is *veiling*: In biographical narrations by the grandparent generation, both Nazi victims and actual perpetrators and their deeds are notably absent. These narrative omissions reflect the actual historical stages in which Jews were dehumanized, persecuted, and exterminated between 1933 and 1945, which leads Rosenthal (1998) to conclude that the real dehumanization and extermination of Jews is psychologically mirrored when both the victims and the crimes of the perpetrators are repressed from consciousness. Rosenthal emphasizes that in both children and grandchildren, repressions that occur psychologically in the grandparent generation manifest themselves as knowledge gaps, which are all too often filled with phantasies about the (grand-)parents' roles in national socialist Germany. Some children and grandchildren even imagine that (grand-) fathers, whose involvements in Nazi crimes are documented in archival records but unknown to their families, were active in the resistance against Nazi Germany (Welzer, Moller, & Tschuggnall, 2002). A second means to avoid responsibility is *victim blaming* or, more generally, a reversal between victims and perpetrators. In secondary antisemitism, for example, Jews or the Allied Forces are blamed for the Holocaust and there is considerable aggression against those who insist on remembering the Holocaust. A third strategy to deflect responsibility is *pseudo-identification with the victims*. Rosenberg cites examples of ostensible philo-Semitism in children of Nazi perpetrators as a strategy of veiling when it goes along with a complete denial of their parents' involvement.

The avoidance strategies of Nazi perpetrators and their families have had consequences far beyond the specific individuals and families involved. First, in Austria, for example, attempts to deflect responsibility determined the postwar political landscape and were so successful that Austria was falsely internationally recognized as the first victim of Nazi Germany (see Uhl, 2001). Second, however, scientific research has mirrored many of these voids: Austrian political officials only ever acknowledged the responsibility and war crimes of Austrians during national socialism starting in 1988; scholarly ignorance mirrored these psychological and larger cultural processes as historical research on Austrian Nazi perpetrators started at about the same time (e.g., Botz, 1987), while before, historians had only devoted attention to resistance by and persecution of Austrians, thus perpetuating the myth that Austrians were victims and not perpetrators of Nazi Germany.

While ignorance can be both socially and psychologically functional for the privileged, powerful, or even perpetrators because it serves to avoid feelings of guilt, not knowing may also fulfill psychological functions for members of subjugated, discriminated, or oppressed groups. North American feminist activists of the late 1960s and 1970s offer a starting point to consider how a position of oppression is translated not into knowledge but into a lack thereof and what psychological mechanisms may be at play here. Like other social movements of the 1960s and 1970s, radical feminists saw liberation as taking place on both an institutional and a psychological plane (see Rosenthal, 1984). In order to connect social and psychological liberation, radical feminists created consciousness-raising as both a political and epistemological method (see Ruck, 2015). Radical feminists engaged with psychology because their theorizing had been incepted by the observation that many women failed to fight against their oppression because they did not realize they were oppressed in the first place. For this reason, radical feminists compiled lists of so-called "resistances to consciousness" (e.g., Peslikis, 1970) that included, for example, glorifying, excusing, or identifying with the oppressor or other privileged groups, over-identifying with one's own oppressed group or other oppressed groups, diverse ways of escapism, overestimating agency in traditionally female roles, individualism, and many others (Sarachild, 1970). As these examples suggest, radical feminists were convinced that there were psychological mechanisms in place that kept members of oppressed groups from gaining insight into their own oppression.

It is critical theory that provided many radical feminists with theoretical direction (e.g., Firestone, 1970) and that more directly tackles the question of not knowing among the oppressed. Starting in the 1920s and against the backdrop of the missing revolution and, later, of rising national socialism in Europe, critical theorists of the Frankfurt School asked why individuals did not revolt against the very conditions they suffered from (see Brunner, Burgermeister, Lohl, Schwietring, & Winter, 2013). They argued, for example, that in late capitalism the nuclear family produced authoritarian personalities who pursued a pseudo-rebellion against social scapegoats instead of revolting against their authoritative father and against those in power. Rising nationalism compensated these authoritarian personalities for their lack of power by affording them the illusion of participating in real power. Bringing these psychoanalytic reflections to bear on gender relations, feminist psychoanalysts have analyzed the formation of femininity under male supremacy. Christa Rohde-Dachser (2003) described the position of many heterosexual women in patriarchal societies as "complementary narcissists": Identified with their fathers, with men in general, and with the male gaze, women subject themselves but at the same time participate in men's successes and power via identification.

In the German-speaking countries, debates about women's psychological oppression and the benefits they gained by association with male privilege culminated in heated arguments about women's so-called "co-perpetration" (in their own oppression) (see Thürmer-Rohr, 2010). On the one hand, this debate revolved around women's roles during Nazi Germany and heavily criticized a long-standing lack of research into women Nazi perpetrators. On the other hand, it was argued that women also partook in the reproduction of patriarchy and other lines of oppression. As Christina Thürmer-Rohr recapitulated, "[w]omen are not only oppressed, abused, and tangled up in a destructive system, they also actively enter this system, win privileges, reap dubious approval, and benefit from their roles insofar as they fulfill them" (2010, p. 89; transl. N.R.).

These analyses help understand why women might be hesitant to give up the range of agency a given set of social relations affords and why, psychologically, they might deny the existence of gender inequities even when faced with them. For example, in the still-masculinist world of science, this may help explain why some accomplished female scientists still adhere to the belief in meritocracy despite battling significant sexism in their fields. Having achieved the approval and respect of their male peers, it may seem self-defeating to threaten this relationship and the privileges it affords by pointing out the sexism in science.

By endorsing meritocracy, the oppressed engage in "not knowing" or at least "un-knowing" the very experiences that keep others like themselves from achieving the same level of success.

Intersectionality helps explain the complex interrelation of oppression and privilege that runs through the above examples. According to intersectionality, various axes of oppression intersect in complex ways on the structural and the psychological level. Both politically and psychologically, these intersections pose the challenge that "those who occupy multiple subordinate identities, particularly women of color, may find themselves caught between the sometimes conflicting agendas of two political constituencies to which they belong" (Cole, 2008, p. 444). Elizabeth Cole has suggested that intersectionality can help move beyond identity politics by drawing the focus to the concrete coalitions individuals and groups build in their attempts to navigate and fight against oppression. Depending on these coalitions or allegiances some systems of oppression might be better knowable than others for those affected by multiple axes of oppression.

For many or even most individuals, oppression intersects with privilege and both positions may go along with not knowing or ignorance about mechanisms and consequences of oppression one either suffers or benefits from. Cole and Zucker (2007) found that US Black women were more likely to identify as feminists than US White women, indicating higher consciousness about gender inequities among Black women. They assumed that experiences of racial oppression sensitized Black women to sexism, too, while Cole (2009) theorized that White women may be complicit with the status quo because as daughters, mothers, or wives of White men they are closer to White male privilege and thus benefit from maintaining racial inequities.

Scholars engaging with the intersections between feminism, critical whiteness studies, and postcolonial studies have coined the term "occidentalist dividend" to understand why White women fend off insights into both their own oppression as women and into the racial and postcolonial order they benefit from (e.g., Dietze, 2010). Public discourse in many European countries, especially the German-speaking countries, has witnessed an obsession with the hijab of Muslim women, which has become almost synonymous with perceptions of women's oppression.

Gabriele Dietze employs the term occidentalist dividend to explain why many women partake in a discourse that problematizes gender inequities especially in Muslim communities and countries but denies them whenever the non-Muslim White majority is concerned. Relying on social psychologist Birgit Rommelspacher, Dietze also claims that "the larger the gulf between pretense [of social equality] and reality the bigger the desire to prove one's own progressiveness via a forced rhetoric of 'emancipation' and liberation" (p. 98). Hence, projecting gender oppression onto Muslim or other "othered" communities is psychologically functional in at least two ways for White non-Muslim women: it allows them to feel liberated, equal to men, and emancipated by contrast with the imagined oppression of Muslim women while at the same time deflecting responsibility and guilt for a system of racial and postcolonial inequality that discriminates against both Muslim men and women.

## The Role of Reflexivity in Disrupting the Psychology of Epistemological Ignorance

In the previous sections we have outlined how psychological processes—such as the operation of identification or of defense mechanisms like projection, repression, or denial—can create and maintain epistemological ignorance among those in positions of power and domination as well as other ways of "not knowing" among those oppressed and subjugated by the dominant group. We might well ask how such psychological processes can be disrupted if the premise of standpoint theory is that one's social location affords the possibility of less partial, more expansive, perspectives. If multiple psychological forces work against becoming aware of and using one's social location as the basis for knowing, how does social location afford opportunities for knowledge, or for "knowing differently"? How can we escape the abyss of not knowing, and even more importantly, not knowing that we do not know? And what are the consequences for a psychology of science and technology?

To start answering these questions we draw on the work of Clare Hemmings, who has argued that the process of translating a social location into critical awareness, knowledge, and even political transformation is mediated through a particular version of reflexivity: a reflexivity marked by affective dissonance (Hemmings, 2012). Hemmings starts her analysis from the position that identity or group characteristics alone cannot suffice as the basis for transformative politics; that simply "being" does not translate into "knowing" or "doing." There is a difference between simply "being" a woman of color in science, for example, and using that ontological status to access situated knowledge and become politicized. She posits that there has to be an affectively unsettling experience of disjuncture between "the experience of oneself over time and the experience of possibilities and limits to how we may act or be" (p. 149). The ability to recognize the gap between ontological and epistemological possibilities is mediated through affect—the rage, frustration, misery, passion, indignation—that is attached to recognizing that one's sense of self (e.g., as a scientist) is not realizable or is thwarted in a system that is fundamentally inequitable (by gender, race, class, etc.). This is not an automatic process. It requires reflexive activity defined as "reflection on the lack of fit between our own sense of being and the world's judgment upon us," a "*negotiation of the difference* between whom one feels oneself to be and the conditions of possibility for a liveable life" (p. 149; emphasis in original). Nonetheless, Hemmings argues that attending to this affective dissonance enables (and might even be required for) generating a counter-episteme that will allow one to know differently, and perhaps then to act differently.

The likelihood that one might experience such affective dissonance is of course influenced by one's position in the social hierarchy and one's relationship to systems of domination and privilege. Hemmings does not unpack or elaborate the conditions that would make it more or less likely that affective dissonance will be experienced and reflexively engaged versus repressed and/or dismissed. She only notes that an affective shift has to occur wherein current conditions are experienced as unfair and an alternative set of possibilities are therefore entertained. As she puts it, "But to move from knowing more to valuing that knowledge requires a shift of some kind…. I suggest that an *affective shift* [emphasis in original] must first occur to produce the struggle that is the basis of alternative standpoint knowledge and politics" (p. 157).

Historically, access to scientific educations and careers and pronouncements about who is suitable for science have been rigidly policed by those in positions of power (see Rutherford, 2015). The experience of affect too has been deemed antithetical to the "scientific attitude" as constructed by white, male, European-descent actors (Keller, 1985). Hemmings' analysis and emphasis on affect suggest that an important part of feminist psychological studies of science must attend to the processes whereby such affective shifts can be encouraged and leveraged as the basis for scientific counter-epistemes.

## Concluding Thoughts on Scientists as (Not) Knowing Subjects

In this chapter, we have asked how epistemological concepts developed by feminist and critical race theorists may be further developed by considering psychological processes to inform a psychology of science that rethinks the relation between scientists' subjectivity, social location, and knowledge or lack thereof. We have pointed out that epistemological ignorance may be psychologically functional for those in power because it allows them to deflect responsibility and ward off guilt. "Resistances to consciousness" among the oppressed or subjugated, on the other hand, may help to imagine oneself as more liberated and emancipated than one actually is, they may serve to avoid conflict with authority figures, and, enable the maintenance of relationships with more privileged individuals and thus the transfer of benefits from these privileges as well.

How are these reflections relevant for psychological studies of science and technology? In the introduction we outlined some of the scholarships that demonstrate how the discipline of psychology has maintained and perpetuated gender stereotypes and how implicit and ingroup bias related to scientists' own gender influences research results and interpretations. More generally, however, systematic ignorance of blind spots and how they relate to one's own position within the matrices of oppression and privilege makes scientists and laypersons alike prone to assume their own experiences, perspectives, and theories as the norm and to reproduce and stabilize power relations. Given that (many) scientists see themselves as immune from such systematic ignorance, perhaps a psychology of science and technology that elucidates the processes through which scientists' own social locations afford or occlude what it is possible to know could help open up new ways of scientific thinking, including about what constitutes science, how to practice it, and whose knowledge is valued.

Standpoint theories emphasize the relation between social location, experience, and knowledge, but they also point out that this relation is not instantaneous but mediated by political struggle and awareness. By drawing on Hemmings (2012) we propose affect as a mediator between social location, experience, and knowledge. In Hemming's view, what is required for the strong objectivity of standpoint epistemology—the translation of a disadvantaged social location into an epistemically advantaged position—is the experience of affective dissonance resulting from a disjuncture between one's sense of one's own capacities and worth and the way one is seen and treated within the social structure. Affect is at the core of Hemming's analysis, precedes the formation of identity, and is perhaps even necessary for the formation of (an activist) identity. However, these experiences of affective dissonance may be epistemologically relevant not only for those in the scientific community that are affected by oppression or social inequality but also for those in positions of power and privilege.

The affective shifts highlighted by Hemmings bear epistemological consequences for scientists. One can easily identify as a woman in science, for example, but identify more strongly with one's male peers and reject the label "feminist" if there has been no experience of affective dissonance and no reflection on that dissonance as the basis for what Hemmings terms "affective solidarity" with other women. Epistemologically, this lack of affective solidarity may well go along with a profound absence of knowledge about gender inequities by the very same subjects who are affected by them. In order to overcome this lack of knowledge the range of emotions (rage, frustration, sadness, misery, passion, indignation) attached to recognizing that one's sense of self is not realizable or thwarted within a fundamentally unequal system need to be acknowledged and lived as a first step. Conversely, it might be other feelings like guilt or shame, that open up the potential of disrupting epistemological ignorance if recognized and experienced by those with power and privileges.

A systematic psychological analysis of how an affective shift occurs (or does not occur) would complement the preceding analysis of how not knowing or epistemological ignorance are enacted and maintained. Such analyses of the psychology and epistemology of affective shifts have to our knowledge not been conducted yet so we here offer some more preliminary suggestions for scientists to exercise reflexivity about their own social positions and their (lack of) experiences as they relate to their own knowledge production. Questions to guide a reflexive analysis might include: Where am I located socially on various dimensions of social inequality or oppression, including gender, sexuality, race/skin color, ethnicity, Nation/state, class, culture, health, age, place of residence/origin, assets, North–South/East–West, social development status (Lutz & Wenning, 2001)? What experiences have I made that are related to my own position as oppressed, discriminated against, subjugated, marginalized, or exploited, what experiences have I made that connect to my own position as privileged, as discriminating against other, as a bystander, as a perpetrator? How have these experiences resonated affectively? Have I ever experienced inequities as ego-dystonic, as an affront to my sense of self or of my values? Under what conditions and what inequities? What axes of social inequities or injustices, if any, have I addressed in my own research? Am I advantaged or disadvantaged, privileged or oppressed on the axes of inequity that I have included in my research?

After we have reflected on these social positions, experiences, and affects as they relate to our own research, we might want to address, individually or collectively with colleagues and/or friends, our very own "resistances to consciousness": Why, specifically, have we not devoted attention to any particular category, always keeping in the back of our minds whether we are privileged in this category or oppressed and that our social position might go along with specific blind spots or "resistances to consciousness" depending on our being privileged or oppressed. How does thinking about one or the other social injustice resonate with me affectively? Do I feel shame, anger, guilt, sadness, rage, frustration, misery, passion, indignation? How do these feelings relate to my research subject or, conversely, to my research voids?

We do realize that this is a rather individualized psychotherapeutic approach to one of the oldest scientific aporias, i.e., the vastness of our ignorance. Nevertheless, we do believe that reflecting upon the affective nature of scientists' subjectivities might be one of the core areas of feminist psychological studies of science and technology. However, beyond attending to the psychological dimensions involved in processes of not knowing, the envisioned feminist psychological studies of science and technology would have to ask, much more systematically than we have done here, how exactly the psychological processes involved in not knowing of ignorance differ according to social location. Can we extrapolate a kind of "psychology of epistemological ignorance" that is specific to privilege and power on the one hand, and a "psychology of not knowing" of those affected by oppression, structural disadvantage, or subjugation, on the other hand? That is, how do power and privilege occlude or actively inhibit what it is possible to know, and conversely, how is "not knowing" also maintained within groups who are oppressed and subjugated? By tackling the affective dimensions of not knowing or ignorance while also being cognizant about their relation to social position and experience, such a feminist psychology of science and technology might finally bring psychology to bear fruit for feminist STS.

## References

Alcoff, L. M. (2007). Epistemologies of ignorance: Three types. In S. Sullivan & N. Tuana (Eds.), *Race and epistemologies of ignorance* (pp. 39–58). Albany, NY: SUNY Press.

Bohan, J. S. (1995). *Re-placing women in psychology: Readings toward a more inclusive history* (2nd ed.). Dubuque, IA: Kendall/Hunt.

Botz, G. (1987). Österreich und die NS-Vergangenheit. Verdrängung, Pflichterfüllung, Geschichtsklitterung. In D. Diner (Ed.), *Ist der Nationalsozialismus Geschichte? Zu Historisierung und Historikerstreit* (pp. 141–152). Frankfurt a.M.: Fischer.

Brunner, M., Burgermeister, N., Lohl, J., Schwietring, M., & Winter, S. (2013). Critical psychoanalytic social psychology in the German speaking countries: A critical review. *Annual Review of Critical Psychology, 10,* 419–468.

Cole, E. R. (2008). Coalitions as a model for intersectionality: From practice to theory. *Sex Roles*, 59(5), 443–453.

Cole, E. R. (2009). Intersectionality and research in psychology. *American Psychologist, 64,* 170–180.

Cole, E. R., & Zucker, A. N. (2007). Black and white women's perspectives on femininity. *Cultural Diversity and Ethnic Minority Psychology, 13,* 1–9.

Collins, P. H. (1986). Learning from the outsider within: The sociological significance of black feminist thought. *Social Problems, 33*(6), 14–32.

Creager, A. N. H., Lunbeck, E., & Schiebinger, L. (Eds.). (2001). *Feminism in twentieth century science, technology, and medicine.* Chicago and London: The University of Chicago Press.

Dietze, G. (2010). 'Occidentalism', European identity and sexual politics. In H. Brundhorst & G. Grözinger (Eds.), *The study of Europe* (pp. 89–116). Baden-Baden: Nomos.

Eagly, A. (2012). Bias, feminism, and the psychology of investigating gender. In R. W. Proctor & E. J. Capaldi (Eds.), *Psychology of science: Implicit and explicit processes.* Oxford University Press. https://doi.org/10.1093/acprof:oso/9780199753628.001.0001.

Firestone, S. (1970). *The dialectic of sex: The case for feminist revolution.* New York: Bantam.

Galton, F. (1869). *Hereditary genius.* London: Macmillan.

Haraway, D. (1989). *Primate visions: Gender, race, and nature in the world of modern science.* New York and London: Routledge.

Harding, S. (1993). Rethinking standpoint epistemology: What is "strong objectivity"? In L. Alcoff & E. Potter (Eds.), *Feminist epistemologies* (pp. 49–82). New York and London: Routledge.

Harding, S. (2004). Introduction: Standpoint theory as a site of political, philosophic, and scientific debate. In S. Harding (Ed.), *The feminist standpoint theory reader* (pp. 1–15). New York and London: Routledge.

Hegarty, P. (2013). *Gentleman's disagreement: Alfred Kinsey, Lewis Terman, and the sexual politics of smart men.* Chicago: The University of Chicago Press.

Held, L., & Rutherford, A. (2012). Can't a mother sing the blues? Postpartum depression and the construction of motherhood in late 20th-century America. *History of Psychology, 15*(2), 107–123.

Hemmings, C. (2012). Affective solidarity: Feminist reflexivity and political transformation. *Feminist Theory, 13,* 147–161.

Hrdy, S. B. (1981). *The woman that never evolved.* Cambridge, MA: Harvard University Press.

Hrdy, S. B. (1999). *Mother nature: A history of mothers, infants, and natural selection.* New York: Pantheon Books.

Hurtado, A. (2010). Multiple lenses: Multicultural feminist theory. In H. Landrine & N. Russo (Eds.), *Handbook of diversity in feminist psychology* (pp. 29–54). New York: Springer.

Keller, E. F. (1978). Gender and science. *Psychoanalysis and Contemporary Thought, 1,* 409–433.

Keller, E. F. (1982). Feminism and science. *Signs: Journal of Women in Culture and Society, 7,* 589–602.

Keller, E. F. (1985). *Reflections on gender and science.* New Haven and London: Yale University Press.

Kumar, N. (2012). Gender and science: Psychological imperatives. In G. J. Feist & M. E. Gorman (Eds.), *Handbook of the psychology of science* (pp. 273–302). New York: Springer.

Lutz, H., & Wenning, N. (2001). Differenzen über Differenz. Einführung in die Debatten. In H. Lutz & N. Wenning (Eds.), *Unterschiedlich verschieden. Differenz in der Erziehungswissenschaft* (pp. 11–24). Opladen: Leske & Budrich.

Martin, E. (1991). The egg and the sperm: How science has constructed a romance based on stereotypical male-female roles. *Signs, 16,* 485–501.

Mayberry, M., Subramaniam, B., & Weasel, L. H. (Eds.). (2001). *Feminist science studies: A new generation.* New York and London: Routledge.

Mendel, I. (2009). Revolution in epistemology? Feminist challenges to epistemic authority in the aftermath of 1968. *Theory and Action, 2*(4), 45–65.

Merchant, C. (1980). *The death of nature: Women, ecology, and the scientific revolution.* San Francisco: HarperCollins.

Mills, C. W. (1997). *The racial contract.* Ithaca, NY: Cornell University Press.

Mills, C. W. (2007). White ignorance. In S. Sullivan & N. Tuana (Eds.), *Race and epistemologies of ignorance* (pp. 13–38). Albany, NY: SUNY Press.

Morawski, J. G. (1985). The measurement of masculinity and femininity: Engendering categorical realities. *Journal of Personality, 53,* 196–223.

Nicholson, I. (2011). "Shocking" masculinity: Stanley Milgram, "obedience to authority", and the "crisis of manhood" in Cold War America. *Isis, 102,* 238–268.

Nicholson, I. A. M. (2001). Giving up maleness: Abraham Maslow, masculinity, and the boundaries of psychology. *History of Psychology, 4,* 79–91.
Parker, I. (2015). Politics and "applied psychology"? Theoretical concepts that question the disciplinary community. *Theory & Psychology, 25*(6), 719–734.
Peslikis, I. (1970). Resistances to consciousness. In S. Firestone & A. Koedt (Eds.), *Notes from the second year* (p. 81). Women's Liberation Movement Collection. Duke University Digital Collections. http://library.duke.edu/digitalcollections/wlmpc_wlmms0139/.
Pettit, M., & Vigor, J. (2015). Pheromones, feminism, and the many lives of menstrual synchrony. *BioSocieties, 10,* 271–294.
Proctor, R. N., & Schiebinger, L. (Eds.). (2008). *Agnotology: The making and unmaking of ignorance.* Stanford, CA: Stanford University Press.
Rohde-Dachser, C. (2003). *Expedition in den dunklen Kontinent. Weiblichkeit im Diskurs der Psychoanalyse.* Gießen: Psychosozial.
Rosenthal, N. B. (1984). Consciousness-raising: From revolution to re-evaluation. *Psychology of Women Quarterly, 8,* 309–326.
Rosenthal, G. (Ed.). (1998). *The Holocaust in three generations: Families of victims and perpetrators of the Nazi regime.* London: Cassell.
Ruck, N. (2015). Liberating minds: Consciousness-raising as a bridge between feminism and psychology in 1970s Canada. *History of Psychology, 18*(3), 297–311.
Ruck, N. (2016). Controversies on evolutionism: On the construction of scientific boundaries in public and internal scientific controversies about evolutionary psychology and sociobiology. *Theory & Psychology, 26*(6), 1–16.
Rutherford, A. (2015). Maintaining masculinity in mid-20th century American psychology: Edwin Boring, scientific eminence, and the "woman problem." *Osiris: Scientific Masculinities, 30,* 250–271.
Rutherford, A. (2017). Surveying rape: Feminist social science and the ontological politics of sexual assault. *History of the Human Sciences, 30,* 100–123.
Rutherford, A., Capdevila, R., Undurti, V., & Palmary, I. (Eds.). (2011). *Handbook of international feminisms: Perspectives on psychology, women, culture, and rights.* New York: Springer SBM.
Rutherford, A., Sheese, K., & Ruck, N. (2015). Feminism and theoretical psychology. In J. Martin, J. Sugarman, & K. L. Slaney (Eds.), *The Wiley handbook of theoretical and philosophical psychology: Methods, approaches, and new directions for social sciences* (pp. 374–392). Hoboken: Wiley.
Sarachild, K. (1970). A program for feminist "consciousness-raising". In S. Firestone & A. Koedt (Eds.), *Notes from the second year* (pp. 78–80).

Women's Liberation Movement Collection. Duke University Digital Collections. http://library.duke.edu/digitalcollections/wlmpc_wlmms01039/.
Scarborough, E., & Furumoto, L. (1987). *Untold lives: The first generation of American women psychologists*. New York: Columbia University Press.
Schiebinger, L. (1993). *Nature's body: Gender in the making of modern science*. Boston: Beacon Press.
Sherif, C. W. (1979). Bias in psychology. In J. A. Sherman & E. T. Beck (Eds.), *The prism of sex: Essays in the sociology of knowledge* (pp. 93–133). Madison: University of Wisconsin Press.
Shields, S. A. (2007). Passionate men, emotional women: Psychology constructs gender difference in the late 19th century. *History of Psychology, 10,* 92–110.
Tuana, N. (2006). The speculum of ignorance: The women's health movement and epistemologies of ignorance. *Hypatia, 21,* 1–19.
Tuana, N. (2008). Coming to understand: Orgasm and the epistemology of ignorance. In R. Proctor & L. Schiebinger (Eds.), *Agnotology: The making and unmaking of ignorance* (pp. 108–145). Stanford, CA: Stanford University Press.
Thürmer-Rohr, C. (2010). Mittäterschaft von Frauen: Die Komplizenschaft mit der Unterdrückung. In R. Becker & B. Kortendiek (Eds.), *Handbuch Frauen- und Geschlechterforschung* (pp. 88–93). Wiesbaden: Springer.
Uhl, H. (2001). Das „erste Opfer": der österreichische Opfermythos und seine Transformation in der Zweiten Republik. *Österreichische Zeitschrift für Politikwissenschaft, 30*(1), 19–34.
Unger, R. K. (1983). Through the looking glass: No wonderland yet! (The reciprocal relationship between methodology and models of reality). *Psychology of Women Quarterly, 8,* 9–32.
Wajcman, J. (2007). From women and technology to gendered technoscience. *Information, Communication & Society, 10,* 287–298.
Welzer, H., Moller, S., & Tschuggnall, K. (2002). *Opa war kein Nazi. Nationalsozialismus und Holocaust im Familiengedächtnis*. Frankfurt a.M.: Fischer-Taschenbuch-Verlag.

# 7

# Social Networks in the History of Psychology

Michael Pettit

In the twenty-first century, the "network" emerged as one of the most prominent ontologies for the social. The widespread adoption of social media platforms such as Myspace, LinkedIn, Facebook, and Twitter have fostered new forms of social organization, what media sociologist danah boyd calls "networked publics" (boyd, 2017). In important respects, these platforms have altered the very nature of psychic life. The Internet has remade human social relations, enabling rapid communication among individuals widely distributed across physical space. The novelty exists. However, our current fascination with new media risks obscuring a longer tradition of network thinking. Part of the concept's appeal is that it operates in various registers, working to bind together these different realms. The network simultaneously serves as a potent metaphor for relationships among individuals, a platform for enacting these relations, and a set of analytic tools for analyzing these

M. Pettit (✉)
Department of Psychology, York University, Toronto, ON, Canada
e-mail: mpettit@yorku.ca



K. C. O'Doherty et al. (eds.), *Psychological Studies of Science and Technology*, Palgrave Studies in the Theory and History of Psychology,
https://doi.org/10.1007/978-3-030-25308-0_7

interactions (Knox, Savage, & Harvey, 2006). For almost a century, social scientists have conceptualized social structures as a network and the tools they developed to study these structures have altered them. Social networks are imagined and real. They are simultaneously material, social, and psychic, braiding together these planes of existence.

This chapter explores what various forms of network analysis can contribute to the reflexive study of psychology as a science. Network thinking draws upon tools developed in mathematics, computer engineering, and sociology, but also represents a venerable tradition within psychology. Indeed, current Social Network Theory (SNA) reflects this historical itinerary through the discipline of psychology and, as a result, has various psychological assumptions built into it. I contend network analysis is fundamentally a psychological approach to social relations, albeit a situationist one where the individual's embeddedness in psychic webs help constitute their personhood. Network analysis can serve an empirical, data-driven endeavor, but also a reflexive one. By reflexivity, I am invoking a large body of scholarship which recognizes that the practice of psychological science is an activity pursued by humans. For this reason, social psychological dynamics govern the behavior of psychologists (Morawski, 2005). Our critical approaches offer what Graham Richards (2002) has called a psychology of psychology. In this chapter, I outline the long history of SNA (rather than focusing on its current vogue), paying particular attention to the contributions of psychologists to sociometry as a precursor to contemporary network science. I then describe certain measures used to formally analyze networks. Finally, I apply this approach to the history of organized psychology in the United States and describe other historical projects which draw upon network analysis.

Networks have come to be everywhere, but this itself is a historical process: the building of a networked society. *The Oxford English Dictionary* offers a helpful semantic history of the expansion of the word "network." In the sixteenth century, it originally applied to woven goods with a structure formed by intersecting lines. By the next century, anatomists began analogizing from these craft goods to the structures of the body. Networks were wet before they were mechanical. The term gained in prominence and usage when it acquired another meaning in the mid-nineteenth century. Then it referred to a "complex system or collection of interrelated things" (Network, 2018a). In other words, network described novel transportation lines and telecommunications routes (cf. Hughes, 1983). In the nineteenth century, scientists drew strong analogies between these mechanical communication systems and the body's nerves (Otis, 2001). Between the end of World War II and 1980, "network" took on a host of new meanings linked to the rise of both mass media and electronic computing. It referred to both the interconnectivity of digital systems and the capacity to broadcast information (Network, 2018a). Perhaps most tellingly, it became a verb (networking) starting in the 1980s. This term appeared in a burgeoning advice literature instructing professionals on how to interact with proximate strangers and acquaintances to advance their careers (Network, 2018b).

A few lessons can be derived from this semantic history. First, the imagery of the lattice is quite old (indeed premodern), one durable and consistent over time. Second, the network came to be understood in the nineteenth century as a *fabric* binding together dispersed elements. This fabric is simultaneously material and symbolic. Finally, networks are tied to communication among these elements; they are understood as unifying people and things into some common whole.

## Networks as Psychological

Given the network's current ubiquity as a metaphor, it is easy to forget the foundational contributions of psychologists to social network analysis (SNA). Kajta Mayer (2012) and Hannah Knox, Mike Savage, and Penny Harvey (2006) offer complimentary genealogies of the network. They demonstrate how social scientists working in this realm braided the technical (the incorporation of graph theory) and the metaphorical (a new vision of interpersonal relations).

Although it is understood as a field at the intersection of sociology, anthropology, and computer science, SNA has a long history within psychology. For example, sociologists Linton C. Freeman and Barry Wellman (1995) credit the Canadian developmental psychologist Helen Bott as a forgotten pioneer. In 1928, she published an ethnography of the play activities among the children she observed at the nursery attached to the University of Toronto. On a daily basis, she selected a different child as a focal point of her observations then systematically counted with whom they talked, interfered, watched, imitated, and cooperated. Freeman and Wellman credit Bott with two innovations later central to network analysis. First, as outlined above, she used what later became known as "focal sampling" in selecting her observations. Second, she organized her data into matrices to depict her participant's interpersonal relationships. Bott was far from an isolated case. Rather she served as a node in a loose network of largely female developmental and educational psychologists who deployed network approaches in the interwar years (Freeman, 1996). However, these women held fairly marginal positions within academia and established no sustained research tradition. They practiced a network thinking, but they failed to receive recognition for developing it as an explicit social theory.

Instead, most histories credit the émigré psychotherapist Jacob Moreno with initiating network analysis as a recognized area of social research (Knox et al., 2006; Freeman, 2004; Mayer, 2012). His 1934 book *Who Will Survive?* popularized these approaches and introduced the term sociometry. Like many post-Freudians, Moreno took an interpersonal (if not outright sociological) approach to psychotherapy as he attempted to get his patients to articulate their place in society. Treatment in Moreno's clinic relied on patients participating in role-playing activities he called psychodramas and charting their interactions with other individuals in network visualizations he called sociograms. In 1937, Moreno launched a *Sociometry*, an interdisciplinary journal dedicated to understanding social structures. Unlike the interwar developmental psychologists, Moreno was programmatic. He conceptualized society itself as a network structure. In introducing sociometry's place among the social sciences, he wrote "Viewing the detailed structure of a community we see the concrete position of every individual in it, also, a nucleus of relations around every individual which is 'thicker' around some individuals, 'thinner' around others. This nucleus of relations is the smallest social structure in a community, a social atom" (1937, 213). For Moreno, these networks were fundamentally psychological in character. Networks were composed of affective bonds or what he called *tele*, "the process which attracts individuals to one another or which repels them, that flow of feeling of which the social atom and the networks are apparently composed" (Moreno, 1937, p. 213). Moreno's therapy entailed examining the patient's various social roles, their affinities with those in their social network, and the tensions arising from these situations. Societies were composed of psychological structures and sociometry was the science which studied them.

Sociometry and its successor SNA played a crucial role in post-war social science. It stressed the embeddedness of individuals and social action. In the 1960s and 1970s, SNA "emerged as probably the most powerful counter to individualistic, rational choice approaches which span the social sciences and are especially important in economics" (Knox et al., 2006, p. 118). In this context, the canonical work was sociologist Mark Granovetter's famous article on "the strength of weak ties" (1973). Looking at information sharing in a community, he found a wide net of loose ties among acquaintances facilitated advancement (e.g., employment) more so than dense personal networks. After Moreno, psychologists continued to make contributions. In the first issue of *Psychology Today*, Stanley Milgram introduced his "small world problem" (Milgram, 1967). He famously calculated how many individual contacts it took for a level to travel between two strangers (one in Nebraska, the other Boston). Finding that "mean number of intermediaries between starters and targets is 5.2," Milgram's study helped popularize the notion of six degrees separating individuals spread across considerable geographical distance (Travers & Milgram, 1969).

## Network Measures

Indeed, the central tenet of SNA is that networks are more than abstract metaphors. However, as Mayer notes, "the sociometrists created their sociograms manually and in an ad hoc fashion" (Mayer, 2012, p. 166). This was certainly true of Moreno and many other psychologists in the sociometric tradition. As late as the 1960s, Milgram recognized how his own approach to networks was pictorial rather than quantitative. However, other researchers were developing more mathematical models representing the data on social interaction. Elaine Forsyth and Leon Katz received credit for conceptualizing the sociogram as a matrix (Forsyth & Katz, 1946), an approach endorsed by Leon Festinger in his empirical work on informal communication networks (Festinger, 1949).

By the late 1970s, the more computationally driven field of SNA overtook sociometry. In 1977, Wellman founded the International Network for Social Network Analysis (INSNA). Two years later, the journal Moreno founded became *Social Psychology Quarterly* and the INSNA supported the creation of a new journal *Social Networks*. In contrast to sociometry, SNA was driven by computational approaches. In the early 1980s, Freeman released UNICET a package of programs for analyzing social networks.

SNA entails a specific vocabulary and a core set of measures that seek to quantify social relations. A node is a unique member of the network. Nodes are connected to one another by an edge or a tie. Every connection carries a numeric value or weight. Multiple connections between individuals increase the weight of their tie. In 1978, Freeman defined three forms of network centrality: degree, betweenness, and closeness. All these measures assess how quickly an individual actor can interact with all others in the network by assessing the shortest pathways across the network. Degree centrality refers to a node's overall connectivity in the network and is obtained by counting the number of surrounding edges. Degree offers an elegant means of conveying an actor's overall visibility within a network. In network analysis, we speak of both simple Degree and Weighted Degree, the latter accounting (and thereby weighing) for multiple connections between two nodes. Betweenness measures centrality in term of a node's place on the path among nodes. This measure assesses whether an actor serves a "brokering" relationship in the network. Individuals with high betweenness centrality may or may not have a high degree of centrality. Finally, closeness centrality accounts for a node that does not share a large number of connections itself, but is connected to such a node. It can serve as a powerful predictor for the flow of communication in an organizational setting.

Even in the age of more computational SNA, psychological assumptions continue to structure the scientific analysis of networks structures. The principle of homophily undergirds the network approach and promises considerable insight into the psychological character of these structures. Homophily holds "similarity builds connection." In other words, there is a remarkable degree of homogeneity in people's interpersonal networks and this commonality functions to bind them together. More socially heterogeneous networks tend toward greater instability, often fragmenting into isolated sub-communities. In the United States, race and ethnicity remain the most salient attributes for network homophily (McPherson et al., 2001). The existence of homophily in formal and informal networks carries significant cognitive consequences relevant to the psychology of science. That like attracts like helps explain why certain shared, unspoken and often unquestioned assumptions tend to coalesce and define disciplinary communities. This homogeneity explains the durability of scientific communities, but also reveals their limits. For example, the overwhelming whiteness of the scientific community often goes unmarked (Morawski, 1997), but acknowledging and naming race and ethnicity as structuring scientific sociality promises considerable insights despite STS's typically colorblind approach to the subject matter (Mascarenhas, 2018).

## Social Networks in the History of Psychology

Network analysis is not only built on insights derived from psychological theory, but it can be used to study psychology as a science. Network approaches are familiar in Science and Technology Studies. In the 1980s, the French anthropologist Bruno Latour famously used the language of networks to describe the power of technoscience, namely its ability to be "so powerful and yet so small" (Latour, 1987, p. 180). Networks like telephone lines or the electric grid were fragile and often invisible but through their connectivity they extended across the globe, becoming everywhere. However, Latour's famous Actor–Network Theory eschewed the formal measures proposed by SNA theorists. In what follows, I will signal the advantages of taking a more formalist approach. Depending on the nature of the project, data for network analysis can come in multiple forms. We can think of born-digital sources generated by our computer-mediated lives, the digital analysis of older corpuses of texts to produce intellectual networks, and finally hand-coding of organizational affiliations to construct networks much like mid-century sociometrists.

Conceptualizing science as a literal network undoubtedly brings to mind the prominence of social media in our current lives. Part of the current vogue for network metaphors is the tractability of the data produced by platforms which facilitates their analysis. This process is what sociologists Noortje Marres and Esther Weltevrede (2013) have called "scraping the social." To return to danah boyd's articulation of "networked publics," she contends these contemporary platforms of sociality possess four properties which distinguish them from the past. First, speech on these platforms is persistent as it leaves permanent, digital traces compared to the more ephemeral nature of the spoken word. Instead of having to record and transcribe it, this kind of speech becomes automatically searchable and retrievable using new computation techniques. This means speech becomes more replicable. One can literally cut and paste speech from social media platforms. Finally, this results in speech becoming consumable by numerous audiences intended and unintended, visible and invisible to the original speaker.

In important respects, contemporary science circulates through these networked publics. The data generated by social media platforms has tremendous potential for understanding scientific practice and communication. Indeed, much scientific activity takes place on these platforms (e.g., the Open Science movement). Furthermore, examining how scientific information moves through different network publics might illuminate certain contemporary controversies (e.g., climate change, GMOs, vaccine hesitancy).

The use of scraped data raises important methodological and ethical concerns. Online data is certainly more stable and searchable than ephemeral speech. However, searching proprietary platforms like Facebook and Twitter is not straightforward. Academic researchers given access to these sites' application programming interface (API) invariably do not receive the entire population of posted data, but rather a sample.

The sampling strategy (the search function) is often proprietary and non-transparent. On an ethical level, data scraped from social media platforms uneasily blurs the distinction between public and private speech. While in many ways, using these data parallels observational research posing minimal risk to participants, it remains unclear whether users provided informed consent for their data to be analyzed when they clicked on terms of service agreements (boyd & Crawford, 2012).

We are not limited to using born-digital sources. The published writings of scientists can be mined to reconstruct their social networks. Historian Derek de Solla Price proposed analyzing the bibliographies of the published article to understand "the nature of the total world network of scientific papers" (Price, 1965, p. 510). Citations to earlier books and articles created links revealing the intellectual and social connections constitutive of scientific communities. This approach was strengthened by co-citation analysis wherein the researcher examines what texts get cited together, revealing ongoing associations among specific scholarly works (Small, 1973). A number of scholars have applied citation analysis to the history of psychology (Davidson, 2018; Fox-Lee, Rutherford, & Pettit, 2016). One can also derive networks from textual sources. For example, Psychologist-historian Christopher D. Green has used linguistic networks to offer novel interpretations of the early history of American psychology. Using full text analysis of key journals such as *Psychological Review* and the *American Journal of Psychology* prior to World War I, he and his collaborators computed the lexical similarity (e.g., word usage) between substantive articles in these journals and clustered the most similar texts together. Their approach has revealed a host of genres of psychological discourse and connected research communities. This extends our understanding of this formative period for psychology, offering a more nuanced view of this era than the Schools approach pitting Functionalism against Structuralism. For example, he detected at least two distinct research communities dedicated to human vision with the one focused on color vision led by Christine Ladd-Franklin (Green et al., 2013).

Finally, one can examine shared membership in organizations. There is a long tradition in the history of science of analyzing correspondence networks to construct a prosopography or a collective biography of members of a scientific community (e.g., Rusnock, 1999; Shapin & Thackray, 1974). Janet Browne famously offered a dramatic reinterpretation of Charles Darwin's career by focusing on how he built his theories through intricate correspondence networks, despite his own social and institutional isolation (Browne, 2002). "Mapping the Republic of Letters" is a digital project hosted by Stanford University which promises a new geography of the Enlightenment by tracing the range and density of different thinkers' reach (e.g., Winterer, 2012).

As these examples suggest, we can enhance and extend the long-standing tradition of prosopgraphy in the history of science by turning to more formal theories and models proposed by SNA. In 1974, sociologist Ronald L. Breiger suggested that examining what he called the duality of persons and groups offered a partial resolution to the person–situation controversy in the social sciences. Reviewing the sociological literature, he argued groups represent the "intersection" of individuals with shared interests. Conversely, much of a person's identity consists in their affiliations with certain groups. For Breiger, the network served as more than a metaphor. Early sociometrists like Moreno focused on interview questions like "who is your best friend?" or "who do you prefer associating with?" In contrast, Breiger proposed a more formal means of generating social networks. He focused on joint membership in (semi-)formal groups (e.g., workplaces, clubs, and other organizations) in constructing his sociometric matrices. Shayna Fox Lee (2014) deployed this approach to investigate the historical geography of disciplinary eminence. Using the entries in *A History of Psychology in Autobiography*, she detailed a number of trends. These include historical shifts in the discipline's self-image as an international versus American field, the stability of Harvard University as a prominent research site across the twentieth century, and the rapid rise of Stanford university in the latter half the century as it came to challenge Harvard's influence. Green and his collaborators (2016) also used this approach to examine the organizational ties between philosophy and psychology circa 1900.

SNA remains a powerful, but underutilized tool for studying the social psychology of science, technology, and innovation. Rather than a general heuristic or a vague metaphor, network measures can help us visualize the group dynamics which sustain science as a cultural activity.

To illustrate the analytic potential of such an approach, I turn to the early history of psychology as an organized and recognized discipline. Namely, I focus on the duality between the first ten presidents of the American Psychological Association (APA) and the institutions where they studied or worked. This tight focus allows me to illustrate the principles of SNA on a small number of individuals. Moreover, this is a familiar topic with a rich historiography. The existence of this scholarship enables a richer interpretation of the graphs.

The creation of professional bodies to demarcate expertise forms a perennial topic in the sociology of professions. At the end of the nineteenth century, American higher education witnessed a tremendous expansion with a host of new institutions established and modeled on the German research university. At this time, psychology was emerging as a recognized discipline, one distinct from physiology on the one hand and philosophy on the other. Gaining recognition revolved around creating institutions unique to psychologists. This included specialized journals, graduate programs, and professional associations. In the United States, the developmental psychologist G. Stanley Hall led the majority of these organizational efforts (Ross, 1972). He served as the leader in two early graduate programs at the newly established universities (first at the Johns Hopkins University and later as the founding president of Clark University). In 1887, he founded the first English-language journal dedicated to psychology (*The American Journal of Psychology*). Finally, in 1892, he convened the first meeting of the APA. Fittingly, he also served as its first president. Hall invited his closest associates to join the APA (Sokal, 1992).

Following the approach outlined by Breiger, I created an array where each row represented an APA president and each column represents a university where they either studied or were employed. In terms of coding, a shared institution does not necessarily mean a concurrently shared institution, although given the time frame this was often the case. This represents a much-simplified version of the history. For illustrative purposes, I only included those institutions where more than one individual studied or taught. For example, Joseph Jastrow spent most of his career at the University of Wisconsin. No other institution loomed larger in his life. However, no other early psychologist shared this affiliation. It created no connection so proved irrelevant for this project. Once data entry was complete, I transposed the array and conducted a matrix multiplication. The resulting matrix shows the institutional ties between psychologists. Because the work presented here is not a study of communication among homogeneous actors, closeness did not seem like a particularly relevant measure. Instead, our analysis focuses on the other two forms of centrality, degree and betweenness.

This graph (Fig. 7.1) tells us some interesting things about the social structure of the early APA. Perhaps the most striking feature is the comparative marginality of William James (because he only possessed a single affiliation, with Harvard). In some ways, this strikes the historian as wrong. James was the most famous psychologist in the United States

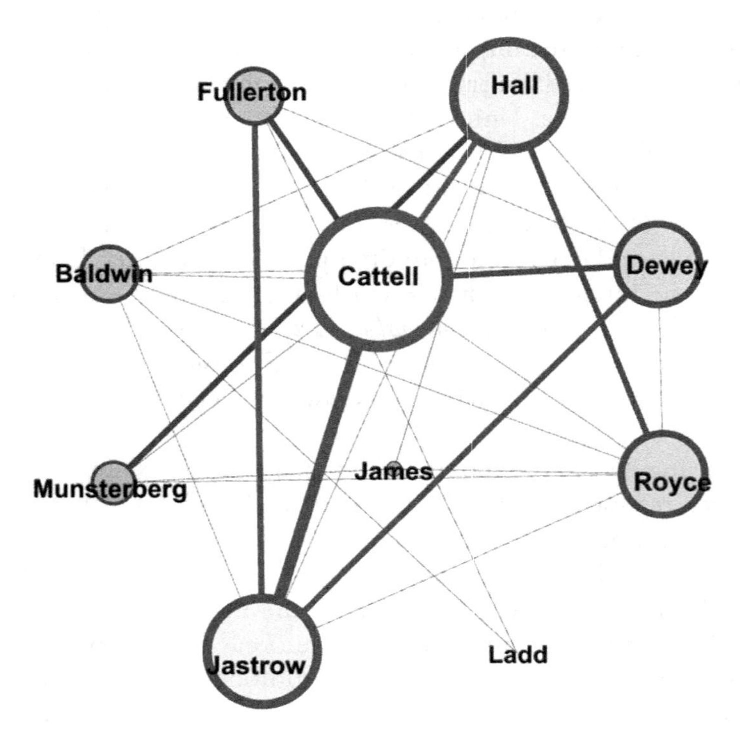


**Fig. 7.1** Sociogram by person of early APA presidents

from the period. His *Principles of Psychology* proved a field-defining textbook. There James articulated ideas about emotion, attention, consciousness, and the self which remain live matters of concerns for psychologists today (Leary, 2018). In important respects, this graph undoubtedly fails to capture the nature and extent of James's considerable influence on psychology and American intellectual history more broadly. However, James's lack of geographic propinquity (in terms of sustained institutional affiliation) with the other founders of the APA had cognitive consequences for the discipline. Historians have long recognized that James's pluralistic vision for psychology did not necessarily mesh well with the men pushing to organize and institutionalize the discipline.

Rather than intellectual originality, the graph captures a different aspect of the history of psychology. In this regard, the greater weighted degree of James McKeen Cattell followed by G. Stanley Hall and Joseph Jastrow is unsurprising. Those psychologists with the highest degree centrality within the early history of the APA exhibit a remarkable degree of homophily, even given the organization's fairly homogeneous character. They epitomized the organization men in science. They created its graduate programs and periodicals. They promoted the new science in a variety of public fora. In other words, the leadership of the early of APA was dominated by men of a practical bent. They were more interested in mental testing than German-style experimental psychology. They helped created a uniquely American approach to psychology: functionalism (Green, 2009) (Fig. 7.2).

The second graph illustrates which institutions loomed largest when it came to becoming an early APA president. What I find most interesting is that though Leipzig (the supposed birthplace of scientific psychology) has a high degree of centrality in the network, it is not the most important place. Instead, having an affiliation with the short-lived philosophy department at the Johns Hopkins University mattered more when it came to becoming an APA president. (For more about the culture of this department see Behrens, 2005; Green, 2007; Pettit, 2013.) Furthermore, Columbia University in New York City became an important home for many of organized psychology's early leaders.

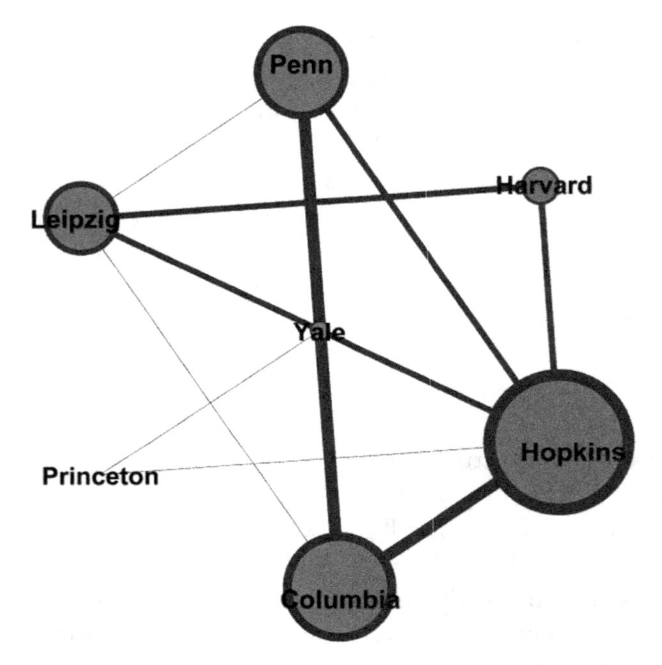


**Fig. 7.2** Sociogram by institution of early APA presidents

# Conclusion

This chapter offered a primer on what sociometry and SNA can contribute to a reflexive psychology of science and technology. Network metaphors are not new to the social study of science. In important respects, they have been there from the beginning, whether in Price's interest in identifying invisible colleges through shared citations or in Latour's more ontologically ambitious mixing of human and nonhuman actors in the making of technoscientific assemblages. Indeed, the focus in this essay on the social as human is somewhat out of step for contemporary Science and Technology Studies. For example, SNA approaches can and have been combined with the insights from Latour's Actor–Network Theory (Pettit, Serykh, & Green, 2015).

However, the approach outlined in this chapter offers more than old wine in new bottles. First, in our current moment, SNA is a potentially useful technology of persuasion. Critical psychologists like Michael Billig (1991) helped dissolve the ancient distinction between rhetoric and (reasonable) thinking. Such work challenges us to consider what argumentative forms are persuasive, why and for whom. The quantitative measures provided by SNA offer a language amenable to other psychological scientists. Although a new psychology of science may wish to eschew the positivist assumptions of its predecessors, I hold quantified measures (understood reflexively and critically) should feature in its argumentative apparatus if it wishes to build bridges to the sciences and incorporate scientists as audiences. Second, concepts like homophily offer novel insight into the psychology of large networks. These forms of associative disposition are not always consciously understood by historical actors. Taken together, the strength of SNA resides in its ability for the psychologist of science to materialize those affective bonds which allowed science as a social endeavor to flourish.

## Works Cited

Behrens, P. J. (2005). The metaphysical club at the Johns Hopkins University (1879–1885). *History of Psychology, 8*(4), 331–346.

Billig, M. (1991). *Ideology and opinions: Studies in rhetorical psychology*. Beverly Hills: Sage.

Bott, H. (1928). Observation of play activities in a nursery school. *Genetic Psychology Monographs, 4*, 44–88.

Boyd, D. (2017, January 16). *Why youth heart social network sites: The role of networked publics in teenage social life*. https://doi.org/10.31219/osf.io/22hq2.

Boyd, D., & Crawford, K. (2012). Critical questions for big data: Provocations for a cultural, technological, and scholarly phenomenon. *Information, Communication & Society, 15*(5), 662–679.

Breiger, R. L. (1974). The duality of persons and groups. *Social Forces, 53*(2), 181–190. https://doi.org/10.1093/sf/53.2.181.

Browne, J. (2002). *Charles Darwin: The power of place*. New York: Knopf.

Davidson, I. J. (2018). The (ab)normal-social-personality catena: Exploring the journal of abnormal and social psychology during the interwar years. *History of Psychology, 21*(2), 151–171.

Festinger, L. (1949). The analysis of sociograms using matrix algebra. *Human Relations, 2*(2), 153–158.

Forsyth, E., & Katz, L. (1946). A matrix approach to the analysis of sociometric data: Preliminary report. *Sociometry, 9*(4), 340–347.

Fox Lee, S., Rutherford, A., & Pettit, M. (2016). II. "Functionalism, Darwinism, and the Psychology of Women" as critical feminist history of psychology: Discourse communities and citation practices. *Feminism & Psychology, 26*(3), 254–271.

Freeman, L. C. (1996). Some antecedents of social network analysis. *Connections, 19*(1), 39–42.

Freeman, L. C. (2004). *The development of social network analysis: A study in the sociology of science*. Vancouver: Empirical Press.

Freeman, L. C., & Wellman, B. (1995). A note on the ancestral Toronto home of social network analysis. *Connections, 18,* 15–19.

Granovetter, M. S. (1973). The strength of weak ties. *American Journal of Sociology, 78,* 1360–1380.

Green, C. D. (2007). Johns Hopkins' first professorship in philosophy: A critical pivot point in the history of American psychology. *American Journal of Psychology, 120,* 303–323.

Green, C. D. (2009). Darwinian theory, functionalism, and the first American psychological revolution. *American Psychologist, 64*(2), 75.

Green, C. D., Feinerer, I., & Burman, J. T. (2013). Beyond the schools of psychology 1: A digital analysis of Psychological Review, 1894–1903. *Journal of the History of the Behavioral Sciences, 49*(2), 167–189.

Green, C. D., Heidari, C., Chiacchia, D., & Martin, S. M. (2016). Bridge over troubled waters? The most "central" members of psychology and philosophy associations CA. 1900. *Journal of the History of the Behavioral Sciences, 52*(3), 279–299.

Hughes, T. P. (1983). *Networks of power: Electrification in Western society, 1880–1930*. Baltimore: Johns Hopkins University Press.

Knox, H., Savage, M., & Harvey, P. (2006). Social networks and the study of relations: Networks as method, metaphor and form. *Economy and Society, 35,* 113–140.

Latour, B. (1987). *Science in action: How to follow scientists and engineers through society*. Cambridge, MA: Harvard University Press.

Leary, D. E. (2018). *Guidebook to James's principles of psychology*. New York: Routledge.

Lee, S. H. A. F. (2014). *Networking Western psychology's elite: A digital analysis of "A history of psychology in autobiography"*. MA thesis, York University, Toronto.

Marres, N., & Weltevrede, E. (2013). Scraping the social? Issues in live social research. *Journal of Cultural Economy, 6*(3), 313–335.

Mascarenhas, M. (2018). White space and dark matter: Prying open the black box of STS. *Science, Technology, & Human Values, 43,* 151–170.

Mayer, K. (2012). Objectifying social structures: Network visualization as means of social optimization. *Theory & Psychology, 22*(2), 162–178.

McPherson, M., Smith-Lovin, L., & Cook, J. M. (2001). Birds of a feather: Homophily in social networks. *Annual Review of Sociology, 27*(1), 415–444.

Milgram, S. (1967). The small-world problem. *Psychology Today, 1,* 61–67.

Morawski, J. G. (1997). White experimenters, white blood, and other white conditions: Locating the psychologist's race. In M. Fine, L. Weis, L. C. Powell, & L. M. Wong (Eds.), *Off white: Readings on race, power, and society* (pp. 13–28). New York: Routledge.

Morawski, J. G. (2005). Reflexivity and the psychologist. *History of the Human Sciences, 18*(4), 77–105.

Moreno, J. L. (1937). Sociometry in relation to other social sciences. *Sociometry, 1,* 206–219.

Network, n. and adj. (2018a). *OED online.* Oxford: Oxford University Press. www.oed.com/view/Entry/126342. Accessed 30 November 2018.

Network, v. (2018b). *OED online.* Oxford: Oxford University Press. www.oed.com/view/Entry/126343. Accessed 30 November 2018.

Otis, L. (2001). *Networking: Communicating with bodies and machines in the nineteenth century.* Ann Arbor: University of Michigan Press.

Pettit, M. (2013). *The science of deception: Psychology and commerce in America.* Chicago: University of Chicago Press.

Price, D. J. D. (1965). Networks of scientific papers: The pattern of bibliographic references indicates the nature of the scientific research front. *Science, 149*(3683), 510–515.

Richards, G. (2002). The psychology of psychology: A historically grounded sketch. *Theory and Psychology, 12*(1), 7–36.

Ross, D. (1972). *G. Stanley Hall: The psychologist as prophet.* Chicago: University of Chicago Press.

Rusnock, A. (1999). Correspondence networks and the Royal Society, 1700–1750. *The British Journal for the History of Science, 32*(2), 155–169.

Shapin, S., & Thackray, A. (1974). Prosopography as a research tool in history of science: The British scientific community 1700–1900. *History of Science, 12*(1), 1–28.
Small, H. (1973). Co-citation in the scientific literature: A new measure of the relationship between two documents. *Journal of the American Society for Information Science, 24*(4), 265–269.
Sokal, M. M. (1990). G. Stanley Hall and the institutional character of psychology at Clark 1889–1920. *Journal of the History of the Behavioral Sciences, 26,* 114–124.
Sokal, M. M. (1992). Origins and early years of the American Psychological Association, 1890–1906. *American Psychologist, 47*(2), 111–122.
Travers, J., & Milgram, S. (1969). An experimental study of the small world problem. *Sociometry, 32,* 425–443.
Winterer, C. (2012). Where is America in the republic of letters? *Modern Intellectual History, 9*(3), 597–623.

# 8

# Engaging Publics on Asthma and Bacteria: Understanding Potential Negative Social Implications of Human Microbiome Research

Amanda Jenkins, Shannon Cunningham and Kieran C. O'Doherty

## Introduction

When new science and technology are developed to address particular health conditions, they do not simply resolve existing problems in a linear fashion. Rather, they emerge within a societal context that itself changes and adapts in often unforeseen ways (Burns, O'Connor, & Stocklmayer, 2003). Novel science and technology thus emerge in the context of pre-existing relationships, vested interests, and institutional

A. Jenkins
University of Guelph, Guelph, ON, Canada
e-mail: ajenki02@uoguelph.ca

S. Cunningham
Alberta Innovates, Edmonton, AB, Canada
e-mail: Shannon.Cunningham@albertainnovates.ca

K. C. O'Doherty (✉)
University of Guelph, Guelph, ON, Canada
e-mail: odohertk@uoguelph.ca





K. C. O'Doherty et al. (eds.), *Psychological Studies of Science and Technology*, Palgrave Studies in the Theory and History of Psychology,
https://doi.org/10.1007/978-3-030-25308-0_8

practices (Wynne, 1992). The uptake and consequences of new knowledge and technologies are therefore not simply a matter of their relative utility and effectiveness, but rather are subject to political, cultural, and institutional arrangements and contingencies. Perhaps more important for the current study, emerging science and technology have social and ethical ramifications beyond the technical purposes they are seen to serve (O'Doherty & Einsiedel, 2012). It is for these reasons that scholars have increasingly called for scientific and technological developments to be accompanied by meaningful public engagement (Collins & Evans, 2007). Such public engagement is not intended to be a marketing of scientific knowledge to the masses, nor is it intended to be a one-directional "polling" of public sentiment used to facilitate translation of new technologies. Rather, meaningful public engagement involves the creation of mechanisms for dialogue, in which publics are introduced to novel areas of science and technology, encouraged to contemplate their implications relative to personal experiences, needs, and values, and are then given the opportunity to articulate their perspectives on the topic. These perspectives can be fed back into scientific and regulatory discourse with the goal of shaping the particular manifestation of technologies (PytlikZillig & Tomkins, 2011).

In this chapter, we consider the role of emerging microbiome science in the context of the lives of people with asthma. In particular, our purpose is to investigate possible social and psychological consequences that may result from biomedical research linking asthma with bacteria. We do this by considering the views of those who are most directly affected by new scientific understandings and medical treatments: individuals who live with asthma and parents of children living with asthma. We begin by presenting an overview of biomedical research linking asthma with bacteria and the human microbiome. We then present an analysis of interviews conducted with individuals who have asthma and parents of children with asthma to understand their perspectives of the potential implications of microbiome research on asthma. We observe, in particular, participants' concerns that associations of asthma with bacteria may inadvertently create negative implications for individuals with asthma. We conclude that care needs to be taken in the translation and dissemination of research linking asthma with bacteria to avoid and/or mitigate such consequences.

## Microbiome Science and Biomedical Research on Asthma

There is a growing interest in examining the microorganisms that live on and in the human body (Wang, Yao, Lv, Ling, & Li, 2017). These microorganisms are collectively referred to as the human microbiome.[1] Research has focused on examining the microbiome in several body sites including the gut, skin, mouth, and vagina. Considered to be an essential part of the human body, the human microbiome has been shown to play an important role in basic biological processes such as regulating the body's immune system (Rees, Bosch, & Douglas, 2018; Ursell et al., 2014; Wang & Li, 2015; Wang et al., 2017). Importantly, perturbations of the microbiome have been linked to the trajectory or development of an increasing number of conditions including asthma (Arrieta et al., 2015), cystic fibrosis (Maughan et al., 2012), inflammatory bowel disease (Kostic, Xavier, & Gevers, 2014), and vaginal health (Albert et al., 2015; Chaban et al., 2014). Asthma, in particular, is a condition which is increasing in prevalence dramatically, with more than 300 million people believed to be affected worldwide (Sullivan, Hunt, MacSharry, & Murphy, 2016). While there is much about asthma that is still unknown (Subbarao, Mandhane, & Sears, 2009), biomedical research has provided new insights into the role that microbes play in the development of asthma (Arrieta et al., 2015; Azad et al., 2013; Couzin-Frankel, 2010; Hahn, 1999; Thomas et al., 2014).

First, human and animal studies suggest that a balance of bacteria and other microbes is important in healthy immune development. Disruption of this balance may lead to the development of diseases including asthma (Couzin-Frankel, 2010). Antibiotics given at a young age may damage the gut microbiome which is critical in the development of the human immune system and thus potentially result in the development of asthma (Arrieta et al., 2015). Studies also suggest that babies born via C-section are more likely to develop asthma

[1]Lederberg and McCray (2001) define the human microbiome as the 'ecological community of commensal, symbiotic, and pathogenic microorganisms that literally share our body space' (p. 8).

than those born vaginally owing to a difference in infant gut bacteria (Azad et al., 2013). Additionally, a mother's use of antibiotics during pregnancy may influence asthma development in early life (Stensballe, Simonsen, Jensen, Bonnelykke, & Bisgaard, 2013). Early antibiotic use may also affect the lung microbiome and future immune responses (Atkinson, 2013). Although microbiome research is relatively young, evidence developed in this field supports the link between antibiotic use, the alteration of microbial ecology and the onset of asthma (Arrieta & Finlay, 2014; Ivanov et al., 2008; Jedrychowski et al., 2011; Russell et al., 2013).

Second, research has also linked the onset of asthma to bacterial infections (Hahn, 1999). This includes acute respiratory infections such as pneumonia, bronchitis, or influenza-like illness (Hahn, 1995). Other evidence has associated chlamydophila pneumonia and mycoplasma pneumonia with new-onset wheezing and decrements in lung function which suggests that these bacterial infections play an important role in the development and severity of asthma (Sutherland & Martin, 2007). Antibiotic therapy has become one means of treatment for asthma developed from bacterial infections which usually has its onset between infancy and 5 years of age. Some research has shown a link between antimicrobial treatment and a reduction of atypical infection and airway inflammation in individuals with asthma (Blasi, Cosentini, Tarsia, & Allegra, 2004). Although antibiotic therapy has been controversial, there is evidence that it can have beneficial effects in reducing asthmatic symptoms (Black, 2007). The term "infectious asthma" is used in some parts of the medical literature to describe asthma believed to be developed from acute respiratory infections such as pneumonia, bronchitis, or influenza-like illness (Hahn, 1995). Given the link between asthma and microbes, the role of antibiotics as a medical intervention relating to asthma is thus somewhat paradoxical: antibiotics may be a key therapeutic tool in resolving or alleviating some forms of asthma, but they are also implicated in damaging the gut microbiome in early childhood in ways that increase subsequent risk of asthma.

In short, there are important implications of microbiome research for individuals who live with asthma (Haw & O'Doherty, 2018). Arrietta et al. (2015) argue that this research can contribute toward the advancement of microbial therapies that prevent individuals from developing asthma. This research also shows promise in understanding the role of the gut microbiome in developing a healthy immune system. Thus, from a biomedical perspective, the positive implications of microbiome research seem relatively uncontroversial. However, existing biomedical literature does not consider potential wider social consequences of emerging microbiome science, nor does it consider the views of those most affected by new knowledge and treatments. Below, we demonstrate a first step to understanding the social and psychological implications of emerging microbiome science, by involving individuals with asthma and parents of children with asthma in conversations about the implications of this new science in their lives.

## Methods

The analysis presented here is part of a larger study investigating experiences of individuals with asthma and parents of children with asthma (see also Haw, Cunningham, & O'Doherty, 2018). This study was approved by the University of Guelph Research Ethics Board. A community sample (n=70) was recruited across Southwestern Ontario. Participants were informed of the study via posters, internet classifieds advertisements, and by other participants. Participants were individuals who had been diagnosed with asthma and/or were the parent of a child with asthma. We used a purposive sampling strategy by first screening potential participants through telephone or email to maximize diversity with respect to education, socioeconomic status, age, sex, place of residence in Ontario, and severity of asthma. Participants provided written consent and received a $20 gift card for their participation. Semi-structured interviews were conducted by SC either in person or over the telephone. Interviews were audio recorded for later transcription and lasted between 45 and 120 minutes. During the interview, participants were introduced to new research linking asthma with a microbial etiology and then asked for their perspectives on potential implications of this research for them. Transcripts were coded thematically by the authors until saturation was reached both inductively and deductively, guided by a focus on concerns explicitly raised by participants and by our research questions. Seventy interviews were conducted over a one-year period (2012–2013). In the transcripts, participant and interviewer are abbreviated as "P" and "I", respectively.

## Analysis

### Negative Associations with Bacteria

We spoke to our participants broadly about research linking bacteria and asthma. Participants and the interviewer together explored the implications of this link. Notable in this exploration was a lack of positive associations with bacteria. In fact, across the entire data set only three examples of positive associations of bacteria were found. In contrast, there are numerous examples of negative associations with bacteria. For example, Participant 47 describes her visceral reaction upon hearing the word bacteria, and her desire not to have asthma connected with bacteria:

> I think…when you say the word bacteria uh I just cringe cause I thought that's the last thing I want to be put with the word asthma.

Participant 47 further described her understanding of how the word bacteria is perceived unfavorably by society:

> P: The first thing that comes to my mind is automatically you cannot use the word bacteria. You cannot use the word bacteria because automatically people are going to hear it and whether there's words that come after that word or not, it's going to stop in people's brains. Um there's gotta be some other word in the medical industry that they can use other than the word bacteria
>
> […] even the word microbe is better than using bacteria because people already have a pre-formed view of what bacteria is and unfortunately, they don't comprehend that there's two types most of the time. So, I mean if you use the word microbes, then that's something that people

> aren't necessarily used to hearing all the time and might be able to be used as the new word for positives, you know? But I- I think the second you say the word bacteria no one's gonna hear what comes out of your mouth after that, they're automatically gonna assume bacteria bad, gross, you know, the illness blah blah blah blah and that's it, because there's been too many bad bacteria's…we were talking about how releasing…and I was sitting there thinkingThinking 'oh man, not that word'.

Participant 47 suggested that the word "bacteria" is associated with dramatic negative representations ("bad, gross"). Due to these negative representations, she expressed very serious reservations about associating asthma with bacteria. The term "microbes", on the other hand, was viewed as less problematic because of a lack of familiarity in public discourse.

Other participants similarly expressed negative reactions toward associating asthma with bacteria, with participants commonly referring to bacteria as germs. The word "germs" was often used by participants in the context of descriptions of a society fearful of bacteria. Participant 28, for instance, described her perspective on broad social perceptions of germs in response to a question around ways to educate individuals about scientific findings relating to a bacterial etiology of asthma:

> P: How should we educate people on this? I don't know. You're gonna have to get a positive outlook on it, but I think that you're right, the probiotic stance and you know, under teaching people that we are germaphobic, I get that everywhere, everywhere we go you know, people are afraid to touch things, like uh constantly sanitizing my hands at work like uh it like it's my job.

Participant 28's description and her use of terms such as "germaphobic" suggest a pervasive and irrational fear of germs across society. Indeed, she implicates herself in this phobia in speaking of her "constantly sanitizing my hands at work". Given these negative reactions to bacteria, many participants spoke strongly against associating asthma with bacteria owing to potential negative social consequences that might follow from this association. We explore this next.

## Consequences of Associating Asthma with Bacteria

The imperative of avoiding the word *bacteria* in connection with asthma was expressed by many participants because of the potential for negative social consequences. In particular, linking asthma with bacteria was a concern to participants because it wrongly implies that asthma is contagious to others. This concern is raised in the following excerpt by Participant 28 in response to being asked to expand on why she believes asthma should not be associated with bacteria:

> Um you know trying to avoid, like bacterial and stuff like, trigger words that are sort of hot buttons in society right now like infectious like bacterial like you know, things like that that you want to avoid because people tend to see that word and then they don't necessarily read all the words around it they just focus in on that and go eww, I'm staying away from that.

Participant 28 suggests that linking asthma with bacteria has negative social consequences because asthma will be viewed as "infectious", here implied to mean contagious. She suggests that the negative associations with bacteria are so strong that people seeing the word might not be sensitive to contextual meanings of particular uses of the term. Furthermore, her description of bacteria as a "trigger word" reflects concerns other participants had around people believing that they can get sick if they come into contact with an individual who has asthma ("ewww, I'm staying away from that"). For some participants, this raised issues around how individuals with asthma manage their illness publicly. For example, Participant 7 describes her negative experience relating to managing her asthma symptoms in public in the following excerpt:

> Like, 'Stay home. Don't spread your germs'. But I'm like 'I don't have germs to spread. It's just [asthma]…'… because we do have this germophobic society where people are, like, constantly scared of getting sick. Um, I definitely think that's a factor. I could see parents, like, not wanting their kids to associate with other kids cause that they think they're sick, but really just have asthma or allergies. I could definitely see that being a factor.

When exhibiting symptoms of asthma (e.g., coughing, wheezing, throat clearing), these are commonly interpreted as symptoms of a communicable illness such as a cold or flu. Participant 7 highlights the social challenges of exhibiting asthma symptoms owing to their similarity to those of communicable diseases. In a "germaphobic society", symptoms attributed to communicable conditions (correctly or incorrectly) are a risk that could lead to people to "not wanting their kids to associate with other kids cause they think they're sick, but really just have asthma or allergies".

Importantly, participants observed that public fears of contagion were associated with bacteria. Participants raised concerns about how representations of the link between asthma and bacteria in health care educational materials could inadvertently stigmatize individuals living with asthma. In particular, Participant 26 discusses in the following exchange how there is a need to differentiate between bacteria that are "contagious" and bacteria that are "not-contagious" in health care education materials to avoid lay public misunderstandings of asthma:

> P: Like just [use] accurate portrayals…it's just a respiratory illness. Like it's nothing contagious, it's not a contagious bacteria. It's just something you have.
>
> R: If you wanted to design…you talked about pamphlets that you've seen. Now if you were to design a pamphlet that was to inform people just about asthma in general, just to let them know what it was like and that, what would you put in it?
>
> P: Information on what it feels like…and the fact that it's not contagious because yeah, like you said, everyone is so concerned with germs and sanitation.

In addition to arguing for accurate portrayals, Participant 26 describes later in his interview how there is a need to be "very careful" when explaining to the lay public how asthma is linked with bacteria. For Participant 26, being "careful" implies here to avoid the suggestion that asthma is linked with a "contagious" form of bacteria, as this could result in increasing public fears around asthma being transmittable. More specifically, participants expressed concerns that linking asthma with a bacterial etiology would lead to further stigmatizing and isolating individuals living with asthma.

Given that many participants were concerned about asthma being viewed as contagious as a result of increasing links between asthma and bacteria, we decided to ask participants during the course of the interview about the term "infectious asthma". None of our participants had heard of the term before the interview, and upon first hearing it almost all of them interpreted the term as referring to a form of asthma that is contagious. When we discussed the intended clinical meaning of the term (i.e., that it refers to asthma that is the result of an infection), participants speculated that the term infectious asthma would similarly be interpreted more broadly as implying asthma that is contagious. Our participants also suggested that any implications that asthma might be contagious would lead to increased stigma of people with asthma. This is illustrated in the following excerpt involving Participant 3 and Participant 4, who are a mother and son who both have asthma:

> P3: Anytime you bring in the word infection they believe it's going to spread like-rapid fire, 'infectious hepatitis'.
>
> P4: When you bring infection into anything, it's automatically stigmatized as being contagious.
>
> P3: No matter, no matter the truth of it.
>
> P4: Yeah.
>
> P3: No matter how solid the truth is behind that.

The word "infectious" is noted by both participants as a term that prevents them from seeing past the pejorative connotations and overshadowing any other context provided about asthma. The comments by Participant 3 and Participant 4 suggest that any kind of associations that are made between asthma and infection in public discourse may lead to difficulties for people with asthma. The word "infection" carries negative connotations that may exacerbate negative portrayals and stigmatization of individuals with asthma. As a result, participants suggested that any association between asthma and bacteria, even if it is part of a medical term like infectious asthma, can lead to negative social consequences for individuals living with asthma. For example, in the following quote,

Participant 53 suggests the terms infection and bacteria should be avoided entirely because these terms could be misconstrued by the lay public and media outlets:

> P: In some ways you have to stay away from the word bacteria and infection and that cause however you explain it to them, the average person is going to twist it that they don't really get the message anyways...I had a quit- quite a bit of experience over the last few years in dealing with the press and what people get and what they don't get. Ah whatever you think they should be getting they don't get.

In addition to concerns that the lay public won't "really get the message", Participant 53 expressed apprehensions in his interview about speaking to the press because they "turn [information] the way they want to turn it". Similar concerns around the uncertainty of how scientific research linking asthma and bacteria would be taken up by the lay public and media outlets were expressed by other participants. This further added to fears some participants described relating to the impact and damaging effects misconstrued information can have around increasing societal beliefs that asthma is transmittable.

## Counteracting Attributions of Contagiousness

Participants in our study described social situations in which they needed to counteract perceptions of contagiousness because of their asthma-related coughing. Many of our participants described using the phrase "it's just asthma" to diffuse social situations in which others treated them as if they were the source of a communicable disease (see also excerpt from Participant 7, above). That is, across the interviews a highly consistent finding was that to counteract attributions of contagiousness for symptoms like coughing, individuals with asthma were able to successfully claim the condition of asthma as one that is not associated with contagiousness ("I don't have germs to spread. It's just [asthma]"). For example, the phrase "it's just asthma" was used by Participant 37, a woman diagnosed with asthma since her early teens, as a way to reduce others' concerns around her coughing:

> P: [If] I'm really coughing I'll be like, 'it's just my asthma like I just, you know I'll say it like if people are looking, 'it's just asthma, like you're fine, like I'm not sick' or if we're visiting people and I'm having coughs and wheezes I'm like, 'I'm not sick it's just my asthma, you know like don't worry you know you're fine, I'm not gonna leave anything'.

Similarly, Participant 63 describes how until she says, "it's just asthma", people are concerned that they may contract a virus from her. In the following excerpt, Participant 63 recalls an experience of needing to explain to the passenger seated next to her on a plane that her coughing was not contagious:

> P: I was on a plane um a short while ago and I I didn't have anything wrong with me I I it was just my asthma. And this lady sat beside me and I coughed um and my lungs were getting a little funny so I took out my inhaler and she looked at me and she said um, she asked if I was uh if I had something contagious because she wanted to move if I did…I thought and I said it's just asthma, I am not contagious, you know don't worry.

Managing the mistaking of symptoms of asthma for symptoms of a cold, in particular, was a common experience of participants. Distinguishing between asthma and a cold was therefore a common task in participants' lives, to reassure others that their coughing was not associated with a cold and therefore not contagious. Participant 2, a mother of a son with asthma, recalls needing to reassure her son who has asthma that he is not contagious despite his brother's persistence that he cover his mouth when coughing:

> P: Yep. Um, I always say to [my son], I always say 'You're not contagious. It's-it's asthma.' And you need to just, like his brother sometimes will say 'Cover your mouth' and Sterling will say 'I'm covering it the best I know how'…But then I have to explain that you know what, it's, it's asth-that's the, that's not a cold cough, that's an asthma cough. And I've gotten to the point where I know the difference between the asthma cough, the cold cough and the allergy cough.

Our analysis suggests that individuals with asthma already experience negative reactions from others who observe asthma symptoms and erroneously attribute these to a contagious condition. The possibility of associating asthma explicitly with a bacterial etiology was viewed by many participants as likely to exacerbate these reactions. Most importantly, it would potentially diminish the rhetorical efficacy of the defense "I'm not contagious, it's just asthma", if asthma came to be seen as contagious.

## Discussion and Conclusion

Our analysis suggests that emerging microbiome science in the context of bacterial etiologies of asthma has the potential to lead to negative implications for individuals with asthma. It is important to emphasize that our analysis is not intended to be a criticism of this science. To the contrary, it is our belief that microbiome science has much to offer both in the treatment and the prevention of asthma and other conditions (O'Doherty et al., 2014). Our main point, therefore, relates to the emergence of this science into a social context characterized by strong aversion to "germs" and the possible or even likely transfer of negative associations from *bacteria* to *asthma*. Our participants' concerns around associating asthma with bacteria is warranted given that previous studies have documented largely negative public perceptions surrounding bacteria. For example, in a study on antibiotic use by Norris et al. (2013), participants expressed a belief that a balance of bacteria was necessary for a healthy body and essential to human survival. Yet, it was noted by participants that widespread advertising of disinfectant products such as cleaning products has led to paranoia about bacteria. In a study on bacterial resistance to antibiotics, Davey, Pagliari, and Hayes (2002) argue similarly that the widespread advertising of antibacterial products, home cleaning agents, and antibiotics has led to strong negative messages in the public about bacteria or germs. This negative perception is described by Davey et al. (2002) as a cultural bias against germs and as leading to the widely held impression that antibiotics are necessary to keep "an overwhelmingly hostile world of bacteria at bay" (p. 44).

These negative public perceptions of bacteria also extend to diseases which are believed to be spread by bacteria. Research has shown that stigma is associated with conditions that are believed to be contagious or harmful to others (Herek, 1999). For example, studies on public perceptions of tuberculosis (TB) suggest that diagnosed individuals are partly stigmatized due to perceived risks of transmission through microbes (Courtwright & Turner, 2010; West, Gadkowski, Ostbye, Piedrahita, & Stout, 2008). This public perception prevails despite medical research suggesting that transmission risk of TB through, for example, airborne microbes from coughing or sneezing, is low unless an individual is exposed to these microbes over a long period of time (American Lung Association, 2016). In addition, according to the World Health Organization (2016) the majority of individuals who are exposed to these microbes do not develop the active disease. The stigmatization of individuals with TB leads to individuals being hesitant to disclose their disease to others for fear of being socially excluded and in some cases being reluctant to adhere to treatment (Dhingra & Khan, 2010). Other diseases associated with stigma because of (incorrectly) perceived transmission risk include disorders such as psoriasis (Halioua et al., 2016) and eczema (Griffiths, Barker, Bleiker, Chalmers, & Creamer, 2016), as well as HIV (Lekas, Siegal, & Leider, 2011) and Hepatitis B (Ellard & Wallace, 2013). These studies support the rationale behind the need many of our participants expressed in having to manage others' perceptions about their symptoms. In particular, the fact that our participants commonly used phrases such as "it's just asthma" to disassociate their symptoms from those of a cold or flu suggests that guarding against attribution of contagiousness is an important aspect of their social management of asthma. It also suggests that if asthma were to become seen as a communicable condition in public discourse, individuals with asthma would lose an important rhetorical resource in the social management of their condition.

Dissemination and uptake of scientific knowledge does not occur in a linear fashion, and it is certainly not possible to predict the precise nature of public understandings of microbiome science and asthma as this enters the public domain. However, our participants' experiences and speculations suggest that there may be unintended negative consequences to this knowledge. An important constraint in our analysis is that participants' statements on the social consequences on microbiome science are speculative. Typically, social scientific methodologies involving surveys, interviews, and focus groups rely on talking with people about topics with which they have intimate familiarity. Conducting an interview study with individuals with asthma and parents of individuals with asthma to learn more about their perspectives about living with asthma is thus not out of the ordinary. However, asking them to comment on the implications of new science and technologies on their lives or on broader society is inherently speculative. This is a potential weakness in studies of this kind and must be taken into account in analysis and interpretation of findings. However, not conducting such studies runs the far greater risk of marginalizing the views of those most affected when scientific and technological advances are integrated into policy frameworks and medical practices. We argue for the importance of conducting studies such as the one we presented in this chapter, while taking care that analytical claims are situated in the context of individuals' speculation about novel science and technology, their experiences of illness and actual and potential stigma, and their experiential knowledge of the health care system into which new biomedical research emerges. For this reason, the concerns they expressed, which are grounded in their everyday experiences of misunderstandings of asthma and stigmatization in a range of life contexts, such as school, work, friendships, and interactions with health care professionals, deserve serious consideration.

It is also important to consider advances in human microbiome research relating to asthma in the context of the success of human microbiome science more broadly. Scientific publications on the human microbiome have increased dramatically over the past few years due to recognition of the potential for microbiome research to transform health care (Slashinski et al., 2013) and lead to important advances in therapies and diagnostics (Gilbert et al., 2016; Haiser et al., 2013; Jia, Li, Zhao, & Nicholson, 2008). This attention from the scientific community has also been taken up in commercial and public domains in the form of companies offering microbiome-based analysis and interventions, as well as heightened media coverage of microbiome related topics (e.g., media coverage of fecal transplants; Chuong, O'Doherty, & Secko, 2015). All of this has led to increased public exposure to alternative discourses about the nature of microbes and their relationship to humans. In particular, these discourses challenge negative pathogen-based perspectives of human illness, and instead offer metaphors of ecosystems and symbiosis to understand the relationship between human and microbes. While our purpose here is neither to endorse nor to challenge these new metaphors (see Juengst, 2009) we do note the potential inherent in these metaphors to counteract negative public associations of bacteria becoming transferred to asthma. Indeed, research on the human microbiome that makes the connection between bacteria and asthma (Arrieta & Finlay, 2014) specifically points to antibiotics and the killing of "healthy bacteria" as the problem. It is this understanding that needs to be leveraged to counteract potential stigmatization of people with asthma. If symptoms of asthma are associated not with pathogens ("bad bacteria"), but rather with past damage to the microbiome ("good bacteria"), many of the fears expressed by our participants may be overcome. There is no easy intervention through which to achieve such positive associations, but if current trends continue and are further augmented by physicians' education on the human microbiome and its implications for health, negative social implications of this new science for people with asthma may be mitigated.

With respect to the larger aim of re-articulating the ways in which psychology can contribute to the study of science and technology, this study illustrates some continuities with previous formulations of a psychology of science, but also important divergences. Similar to previous articulations of a psychology of science (e.g., Feist, 2006) this study recognizes the social aspects that accompany the development and application of scientific knowledge. However, in contrast to such attempts, this study goes beyond a characterization of human phenomena in terms of variables (O'Doherty & Winston, 2014). The study integrates principles of *scientific realism* and *social constructionism* in its analysis. In particular, our study relies on the premise that the biological foundations of conditions such as asthma and knowledge about microbes and their relationship with asthma can be usefully described in the language of the natural sciences. At the same time, we see scientific knowledge as embedded within larger social relations and emerging within particular social contexts. As such, our study orients to the possible trajectories of meaning that may emerge and develop in relation to this field of science and how this may affect people. While we certainly do not claim that all studies that purport to instantiate a psychological study of science and technology need to take such an approach, we do believe that any framework that does not allow for such an approach is at best incomplete or, at worst, flawed.

## References

Albert, A. Y., Chaban, B., Wagner, E. C., Schellenberg, J. J., Links, M. G., Van Schalkwyk, J., ... & VOGUE Research Group. (2015). A study of the vaginal microbiome in healthy Canadian women utilizing cpn60-based molecular profiling reveals distinct Gardnerella subgroup community state types. *PLoS One, 10*(8), e0135620.

American Lung Association. (2016). *Learn about tuberculosis*. Retrieved from http://www.lung.org/lung-health-and-diseases/lung-disease-lookup/tuberculosis/learn-about-tuberculosis.html.

Arrieta, M. C., & Finlay, B. (2014). The intestinal microbiota and allergic asthma. *Journal of Infection, 69*(1), 53–55.

Arrieta, M. C., Stiemsma, L. T., Dimitriu, P. A., Thorson, L., Russell, S., Yurist-Doutsch, S., ... Brett Finlay, B. (2015). Early infancy microbial and metabolic alterations affect risk of childhood asthma. *Science Translational Medicine, 7*(307), 1–14. https://doi.org/10.1126/scitranslmed.aab2271.

Atkinson, T. P. (2013). Is asthma an infectious disease? New evidence. *Current Allergy and Asthma Reports, 13*(6), 702–709.

Azad, M. B., Konya, T., Maughan, H., Guttman, D. S., Field, C. J., Chari, R. S., ... Kozyrskyj, A. L. (2013). Gut microbiota of healthy Canadian infants: Profiles by mode of delivery and infant diet at 4 months. *Canadian Medical Association Journal, 185*(5), 385–394. https://doi.org/10.1503/cmaj.121189.

Black, P. N. (2007). Antibiotics for the treatment of asthma. *Science Direct, 7,* 266–271.

Blasi, F., Cosentini, R., Tarsia, P., & Allegra, L. (2004). Potential role of antibiotics in the treatment of asthma. *Current Drug Targets, 3*(3), 237–242.

Burns, T. W., O'Connor, D. J., & Stocklmayer, S. M. (2003). Scientific communication: A contemporary definition. *Public Understanding of Science, 12,* 183–202.
Chaban, B., Links, M. G., Jayaprakash, T. P., Wagner, E. C., Bourque, D. K., Lohn, Z., … & Hill, J. E. (2014). Characterization of the vaginal microbiota of healthy Canadian women through the menstrual cycle. *Microbiome, 2*(1), 23.
Chuong, K. H., O'Doherty, K. C., & Secko, D. M. (2015). Media discourse on the social acceptability of fecal transplants. *Qualitative Health Research, 25*(10), 1359–1371.
Collins, H., & Evans, R. (2007). *Rethinking expertise*. Chicago, MI: University of Chicago Press.
Courtwright, A., & Turner, A. N. (2010). Tuberculosis and stigmatization: Pathways and interventions. *Public Health Reports, 125*(4), 34–42.
Couzin-Frankel, J. (2010). Bacteria and asthma: Untangling the links. *Science, 330*(6008), 1168–1169.
Davey, P., Pagliari, C., & Hayes, A. (2002). The patient's role in the spread and control of bacterial resistance antibiotics. *Clinical Microbiology & Infection, 8*(2), 43–68.
Dhingra, V. K., & Khan, S. (2010). A sociological study on stigma among TB patients in Delhi. *The Indian Journal of Tuberculosis, 57*(1), 12–18.
Ellard, J., & Wallace, J. (2013). *Stigma, discrimination, and hepatitis B.* Melbourne: Australian Research Centre in Sex, Health and Society, La Trobe University.
Feist, G. J. (2006). *The psychology of science and the origins of the scientific mind.* New Haven and London: Yale University Press.
Gilbert, J. A., Quinn, R. A., Debelius, J., Xu, Z. Z., Morton, J., Garg, N.,…& Knight, R. (2016). Microbiome-wide association studies link dynamic microbial consortia to disease. *Nature, 535*(7610), 94–103.
Griffiths, C. E. M., Barker, J., Bleiker, T., Chalmers, R., & Creamer, D. (Eds.). (2016). *Rook's textbook of dermatology* (Vol. 4). Chichester, West Sussex: Wiley.
Hahn, D. L. (1995). Infectious asthma: A reemerging clinical entity? *The Journal of Family Practice, 41*(2), 153–157.
Hahn, D. L. (1999). Chlamydia pneumoniae, asthma, and COPD: What is evidence? *Annals of Allergy, Asthma, & Immunology, 83*(4), 271–288.
Haiser, H. J., Gootenberg, D. B., Chatman, K., Sirasani, G., Balskus, E.P., & Turnbaugh, P. J. (2013). Predicting and manipulating cardiac drug inactivation by the human gut bacterium *Eggerthella lenta. Science, 341*, 295–298.

Halioua, B., Sid-Mohand, D., Roussel, M. E., Maury-le-Breton, A., de Fontaubert, A., & Stalder, J. F. (2016). Extent of misconceptions, negative prejudices, and discriminatory behaviour to psoriasis patients in France. *Journal of the European Academy of Dermatology and Venereology, 30*(4), 650–654.

Haw, J., Cunningham, S., O'Doherty K. C. (2018). Epistemic tensions between people living with asthma and healthcare professionals in clinical encounters. *Social Science & Medicine, 208,* 34–40. https://doi.org/10.1016/j.socscimed.2018.04.054.

Haw, J., & O'Doherty, K. (2018). Clinicians' views and expectations of human microbiome science on asthma and its translations. *New Genetics and Society, 37*(1), 67–87. https://doi.org/10.1080/14636778.2018.1430561.

Herek, G. (1999). Aids and stigma. *American Behavioral Scientist, 42*(7), 1106–1116.

Ivanov, I. I., de Llanos Frutos, R., Manel, N., Yoshinaga, K., Rikin, D. B., Sartor, R. B., ... Littman, D. R. (2008). Specific microbial flora direct the differentiation of Th17 cells in the mucosa of the small intestine. *Cell Host Microbe, 4*(4), 337–349. https://doi.org/10.1016/j.chom.2008.09.009.

Jedrychowski, W., Perera, F., Maugeri, U., Mroz, E., Flak, E., Perzanowski, M., & Majewska, R. (2011). Wheezing and asthma may be enhanced by broad spectrum antibiotics used in early childhood: Concept and results of a pharmacoepidemiology study. *Journal of Physiology and Pharmacology, 62*(2), 189–195.

Jia, W., Li, H., Zhao, L., & Nicholson, J. K. (2008). Gut microbiota: A potential new territory for drug targeting. *Nature Reviews, 7,* 123–129.

Juengst, E. T. (2009). Metagenomic metaphors: New images of the human from 'translational' genomic research. In M. A. M. Drenthen, F. W. J Keulartz, & J. Proctor (Eds.), *New visions of nature: Complexity and authenticity* (pp. 129–145). New York, NY: Springer.

Kostic, A. D., Xavier, R. J., & Gevers, D. (2014). The microbiome in inflammatory bowel disease: Current status and the future ahead. *Gastroenterology, 146*(6), 1489–1499.

Lederberg, J., & McCray, A. T. (2001). 'Ome sweet' omics—A genealogical treasury of words. *Scientist, 15*(7), 8.

Lekas, H. M., Siegal, K., & Leider, J. (2011). Felt and enacted stigma among HIV/HCV-coinfected adults: The impact of stigma layering. *Qualitative Health Research, 21,* 1205–1219.

Maughan, H., Wang, P. W., Caballero, J. D., Fung, P., Gong, Y., Donaldson, S. L., … & Waters, V. J. (2012). Analysis of the cystic fibrosis lung microbiota via serial Illumina sequencing of bacterial 16S rRNA hypervariable regions. *PLoS One, 7*(10), e45791.

Norris, P., Chamberlain, K., Dew, K., Gabe, J., Hodgetts, G., & Madden, H. (2013). Public beliefs about antibiotics, infection and resistance: A qualitative study. *Antibiotics, 2*, 465–476.

O'Doherty, K., & Einsiedel, E. (Eds.). (2012). *Public engagement and emerging technologies*. Vancouver: University of British Columbia Press.

O'Doherty, K., Neufeld, J. D., Brinkman, F. S. L., Gardner, H., Guttman, D. S., & Beiko, R. G. (2014). Conservation and stewardship of the human microbiome. *Proceedings of the National Academy of Sciences, 111*(40), 14312–14313. https://doi.org/10.1073/pnas.1413200111.

O'Doherty K., & Winston A. (2014). Variable, overview. In T. Teo (Ed.), Encyclopedia of critical psychology. SpringerReference (www.springerreference.com). Springer-Verlag Berlin Heidelberg. https://doi.org/10.1007/SpringerReference_3489842013-02-0108:28:43UTC.

PytlikZillig, L. M., & Tomkins, A. J. (2011). Public engagement for informing science and technology policy: What do we know, what do we need to know, and how will we get there? *Review of Policy Research, 28*, 197–217. https://doi.org/10.1111/j.1541-1338.2011.00489.x.

Rees, T., Bosch, T., & Douglas, A. E. (2018). How the microbiome challenges our concept of self. *PLoS Biology, 16*(2), e2005358.

Russell, S. L., Gold, M. J., Willing, B. P., Thorson, L., McNagny, K. M., & Finlay, B. B. (2013). Perinatal antibiotic treatment affects murine microbiota, immune responses and allergic asthma. *Gut Microbes, 4*(2), 158–164.

Slashinski, M. J., Whitney, S. N., Achenbaum, L. S., Keitel, W. A., McCurdy, S. A., & McGuire, A. L. (2013). Investigators' perspectives on translating human microbiome research into clinical practice. *Public Health Genomics, 16*(3), 127–133. https://doi.org/10.1159/000350308.

Stensballe, L. G., Simonsen, J., Jensen, S. M., Bonnelykke, K., & Bisgaard, H. (2013). Use of antibiotics during pregnancy increases the risk of asthma in early childhood. *The Journal of Pediatrics, 162*(4), 832–838.

Subbarao, P., Mandhane, P. J., & Sears, M. R. (2009). Asthma: Epidemiology, etiology and risk factors. *Canadian Medical Association Journal, 181*(9), 181–190.

Sullivan, A., Hunt, E., MacSharry, J., & Murphy, D. M. (2016). The microbiome and the pathophysiology of asthma. *Respiratory Research, 17*(163). https://doi.org/10.1186/s12931-016-0479-4.

Sutherland, E. R., & Martin, R. J. (2007). Asthma and atypical bacterial infection. *Chest, 132*(6), 1962–1966.

Thomas, L. T., Arrieta, M.-C., Dimitriu, P. A., Thorson, L., Mohn, W. W., Finlay, B. B., et al. (2014). The impact of the intestinal microbiome on human immune development and atopic disease. *Allergy, Asthma, and Clinical Immunology, 10*(1), A63. https://doi.org/10.1186/1710-1492-10-s1-a63.

Ursell, L. K., Haiser, H. J., Van Treuren, W., Garg, N., Reddivari, L., Vanamala, J., ... & Knight, R. (2014). The intestinal metabolome: An intersection between microbiota and host. *Gastroenterology, 146*(6), 1470–1476.

Wang, B., & Li, L. (2015). Who determines the outcomes of HBV exposure? *Trends in Microbiology, 23*(6), 328–329.

Wang, B., Yao, M., Lv, L., Ling, Z., & Li, L. (2017). The human microbiota in health and disease. *Engineering, 3*(1), 71–82.

West, E. L., Gadkowski, L. B., Ostbye, T., Piedrahita, C., & Stout, J. E. (2008). Tuberculosis knowledge, attitudes, and beliefs among North Carolinians at increased risk of infection. *North Carolina Medical Journal, 69*(1), 14–20.

World Health Organization. (2016). *Fact sheet on tuberculosis*. Retrieved from http://www.who.int/3by5/TBfactsheet.pdf.

Wynne, B. (1992). Misunderstood misunderstanding: Social identities and public uptake of science. *Public Understanding of Science, 1,* 281–304.

# Part III

## Critical Perspectives on Psychology as a Science

# 9

# A New Psychology for a New Society: How Psychology Can Profit from Science and Technology Studies

Estrid Sørensen

Psychology is not only an important extension to social studies of science. It is itself a scientific discipline and thus a potential object of science studies. How well is psychology equipped with vocabulary and methods to examine itself as an object of science studies? I discuss this in relation to the concern that psychology currently may not meet the challenges of contemporary society, and that it may not relate adequately to the world of phenomena it investigates in its attempt to meet these challenges (cf. Huniche and Sørensen, 2019a, b). In introducing these discussions I begin with a brief look back to the first formation of psychology as a discipline.

The introduction of official recording and documentation of crime rates in the 1830s and of unemployment rates around 1900 shaped the image of an individual who could be singled out and compared with other individuals (Porter, 1995). Such figures made populations knowable and governable, and thus predictable as a mass. The invention of

E. Sørensen (✉)
Department of Social Science, Ruhr-University Bochum,
Bochum, Germany
e-mail: estrid.sorensen@rub.de





K. C. O'Doherty et al. (eds.), *Psychological Studies of Science and Technology*, Palgrave Studies in the Theory and History of Psychology,
https://doi.org/10.1007/978-3-030-25308-0_9

the notion of an individual allegedly freed from tradition (Asplund, 1985; Jensen & Sørensen, 1995) also came to provide the moderns with the psychological task of shaping and governing their own identities (Rose, 1999). Along with the other social sciences, psychology developed as an academic discipline over a period in which questions about how to organise people's social positions, and how to shape identities, were of crucial concern. Social migration, new forms of governance, and changes in schemes of social inclusion and exclusion were pressing problems. The social sciences emerged to address these problems of modern society, which thus came to shape the "sociological gaze" (Deleuze, 1979). Psychology served both to provide predictive knowledge of individuals, which was helpful for setting norms and governing populations, and also as a science that provided individuals with vocabularies and methods for understanding themselves. The latter was helpful both for self-governance in accordance with existing norms, and with a view to developing methods to emancipate from such norms (cf. Porter, 1995). The dominating imaginary framing modern politics, engineering and science—including psychology—was a relatively homogenous society with a general consensus about belonging to a shared national culture. In such a culture, differences and their management would be purified, separated, and organised in (again relatively homogenous) professions, disciplines, unions, classes, families, ages, gender, ethnic groups, etc., in which locations and times would be segregated into different functions and the threats and sources of danger would be identifiable. In a situation dominated by the named social problems and the general idea of homogenous and well-settled structures being necessary for their solution, knowledge regimes that provided information and predictive understanding about such individuals, groups, times, spaces and dangers were helpful and needed. This was for the social (re)production of modern society thus organised, for managing such groups, spaces, times and dangers, and for the ability of individuals to reflect on, critique and move between the groups, spaces and times, and to manage the dangers.

Among many other authors of social science and observers of contemporary society, Ulrich Beck (1992) emphasised in his seminal book *Risk Society* that the core challenges of the contemporary and future world will be the unintended effects of modernisation, such as ecological catastrophes (pollution, droughts, etc.) and technological disasters (chemical leaks, nuclear accidents, etc.). Such challenges affect lives across the boundaries within modern society and disrespect the separation of functionally structured societal groups, times and spaces. They create vulnerabilities, mobility and potentially also solidarities across what in modern society are otherwise distinct social groups (Beck, 2016). Not only catastrophes and disasters, but many mundane contemporary concerns around scientific, technological, social and natural developments—such as AI, pre-implementation diagnostics, face recognition, waste separation and forest fires—contest the boundaries between the social, the natural and the technological. This makes it increasingly difficult, if not impossible, to point to one single discipline, profession, vocabulary and/or method that can address such concerns and intervene in the both technical, social and natural characteristics of the emerging and changing concerns. Interdisciplinary, interprofessional, cross-theoretical and multi-methodological collaborations are needed. A distinct and homogenous societal division of labour does not meet this requirement. Our contemporary and future societies have a lesser requirement for science—and psychology—to make populations and individuals knowable, predictable and governable as homogenous and separate entities; instead these societies are more in need of (psychological) knowledge and methods for engaging with differences, surprises and change, and for learning to collaborate across disciplines, discourses and diversities.

While other disciplines—anthropology and sociology in particular—have dealt with these challenges for a long time, psychology seems to be lagging behind and sticking to its well-established modern gaze. Psychological knowledge of the modern kind is surely still needed, and likely to keep psychology busy for some time to come. However, in this chapter I will present the concern that in order to respond to contemporary challenges and contribute to their resolution, psychology needs to reinvent itself. I argue that Science and Technology Studies (STS) offer vocabularies that may help scientific psychology both to better reflect on its own relationship with the world and its contemporary challenges, and to adapt to and engage with these challenges.

With reference to a concept coined by Helen Verran (2001), the chapter begins by characterising psychology as a *distant judging observer*. I argue that psychology's vocabulary does not serve it well to reflect upon its own practices and relationship with the world. Psychology, I continue, needs not only to reorganise this relationship, but also to equip itself with a better reflexive vocabulary to reflect on it. I finally present three core concepts of STS which I argue can enable psychology to be more self-reflexive, and improve the way it adapts to and engages with the challenges of our contemporary and future world.

## Psychology as a Distant Judging Observer

How can one talk about psychology in the singular? One of the challenges to psychology as it formed as a discipline was due to its discrepancies and disunity. According to Theodor Porter (1995), it was these characteristics that drew psychology towards quantification. The shallow theorising required in quantitative procedures worked as a device that allowed scholars across different conceptual provinces to work together, and to form a discipline. As is the case with all disciplines (cf. Barry & Born, 2013), psychology has always been heterogeneous and a home for different perspectives and tensions. Nonetheless, the discipline talks about itself as *a* discipline—in the singular—and while it is indeed useful to be able to talk about psychology in the singular, it is important to keep in mind the generalisation involved in doing so, and that such generalisation always comes at the cost of attention being paid to differences that obviously remain. In my discussion of quantitative psychology, I hope to make clear that even though considerable variations exist within this area of psychology, it is also possible to identify common patterns.

In quantitative psychology, which remains the largest, best funded and most publicly visible area of the discipline, numbers and statistics make up the concepts and language in which psychological phenomena are expressed. It is a binary language of more or less, higher or lower, wider or narrower, etc. Scales have two directions and accordingly, quantification has come to characterize psychological phenomena as binary, or indeed, as a combination of many binaries. While quantification has the amazing capability of describing any phenomenon, it sets important limits on what it can possibly mean to know something. Religious and metaphysical insights, collective memories, common sense, practical and material familiarity, natural experiences, intuitions, implicit, non-verbalised and bodily knowledge, gut feelings, sensations, emotions, etc. are all excluded from the realm of knowing in quantitative psychology. Or put differently: most of our mundane everyday cognition, our judgements and our basis for understanding cannot be acknowledged as relevant knowledge in quantitative psychology. At most, non-quantitative insights can be considered material for producing quantitative data as a foundation for scientific insight. Even though other areas of psychology, which often apply non-standardised qualitative methods, will also translate mundane everyday insights into scientific language, such insights can and often are acknowledged as insights that are useful for scientific theorising, not only as data for generating scientific insights.

Quantitative psychology's implicit differentiation between relevant quantitative and irrelevant non-quantitative modes of knowing and explaining does not only classify knowledge, but also defines the relationship between the knower and the known. By excluding various forms of knowing, quantitative psychology defines these as "Other" to its own form of knowledge production. These othered modes of knowing thus come to be in need of explanation (i.e. quantification) while the psychologist's quantification needs no further explanation (other than the discussions that form the foundations of quantitative psychology). Quantitative psychology establishes a constellation of two homogenous regions of both knowledge and knowers (cf. Sørensen, 2007, 2008): one is inhabited by knowers whose modes of knowing are questionable and in need of explanation; the other is inhabited by knowers whose mode of knowing may indeed be discussed in detail, but as a whole is already explained and granted recognition. In establishing this constellation, a division of labour is also generated that allocates to knowers of the latter region the task of explaining the knowledge of the former and the competencies to do so, while those in the former region must await explanation and lack the competencies to explain themselves. Furthermore, the language—i.e. numerical language—that quantitative psychology sets as the gold standard to represent the psychological phenomena belongs only to one region, and therefore the use of it reinforces its difference from the Other region where there are Other ways of talking and thus thinking about psychological phenomena and about ourselves.

This regional separation and homogenising of different modes of knowing that belong to separate groups resonates with the purified constellation of homogenous, well-settled and separate groups, times, spaces and dangers, which I described in the introduction as characteristic of the organising principle of modern society. This principle organised both the establishing of modern society, the solving of its problems and the generation of its knowledge. There is no doubt that quantitative psychology served modern society well, and still serves large parts of contemporary society that keep following this organising principle. However, as I also discussed in the introduction, different ways of organising society are already well developed and they even seem to be expanding. We may talk about these as having an entangled character, compared to the regional and purifying character of modernist organising principles. Characteristic of the entangled organising principle are the emerging processes of connecting, of collaborating, and of entangling heterogeneous actors and components. The strong reduction of their concept of scientific knowledge production to only a very narrow set of methods makes it difficult for quantitative psychology's modes of knowing to engage with difference without doing away with it (cf. Verran, 1999) and it constrains this area of psychology's ability to relate to and interact with other modes of knowing. Quantitative psychological knowledge can only relate to other modes of knowing by translating them into terms relatable to quantification (such as better and worse, more or less, etc.) and thus altering those accordingly that did not originally have this form. This is the case not only for: (a) mundane everyday forms of knowledge production, but very much too for (b) other areas of psychology, other disciplines such as history, plus large areas of social science and the humanities, and also for (c) modes of knowing that involve non-Western social and technical practices. Again, with reference to my brief account of contemporary problems and societal needs, these deficits must be regarded as severe and substantial. It is the constellation of regions delimited by boundaries and inhabited by homogenous knowers that I attempt to challenge in this chapter, and which I am concerned stands in the way of psychology's ability to reflect on itself and to confront the challenges of our contemporary world.

## The Inadequacy of Quantitative Psychology to Reflect on Psychology's Practice

I have argued throughout this paper that psychology needs to reinvent itself to be able to meet the challenges of contemporary and future society. Methods for reflecting on one's own practice are urgently needed when a person, an organisation or a discipline needs to change or develop (e.g., Schön, 1983). Particularly in recent years, quantitative psychology has increasingly applied its methods to reflect on its own scientific practices (e.g., Brown & Heathers, 2016; Colin et al., 2018; Etz & Vandekerckhove, 2016). However, numerous figures in STS (Beck, Niewöhner, & Sørensen, 2012; Hess, 1997; Sismondo, 2004) argue that quantitative science's ability to reflect on its own practices is limited, since it is necessarily based on *normative* ideas about how science is conducted, rightly or wrongly. In psychological experiments, for instance, hypotheses have to be generated prior to the empirical study, variables should be controlled, measurements should be objective and results should state insights about cause and effect relationships. STS scholars describe science studies that build their empirical investigations on such principles as normative. This is because a standard way in which scientific knowledge should be done is determined prior to empirical studies, and accordingly the focus of the empirical study is on whether the investigated scientific practice complies to these principles, or how to better do so (cf. Bloor, 1991). Such normative principles, however, do not provide psychological scientists with operational methods to query their own practices, i.e. what they actually do, how they do it and what reasons (good, bad or simply pragmatic) there might be for doing so, independently of whether this conforms to pre-set principles.

This is a shortcoming because it creates fictions about these principles guiding scientific practices (cf. Mulkay & Gilbert, 1981); hinders analysis and discussions of other principles that in practice may be even more influential on scientific conduct; and limits debates about whether these principles are indeed the most relevant for how to produce sound scientific knowledge (cf. Derksen, 2019).

This shortcoming can be observed clearly in the current debates around the "crisis of confidence" (Pashler & Wagenmakers, 2012) in psychological science. After the discovery of a series of serious scientific malpractices in the discipline (Stroebe, Postmes, & Spears, 2012), scholars began reflecting upon possible shortcomings in psychological science, and launched several initiatives to remedy them. One of these concerned replication. Thinking with standardised principles of quantitative psychology, scientists lamented the general lack of replication of psychological experiments (Baker, 2015), and argued that if experiments would only be replicated, malpractice could be discovered earlier. Furthermore, the general level of confidence, which decreased particularly within US-American psychology, would be restored if people could see that results in psychological science are indeed systematic. Therefore, initiatives were taken to increase replication of studies (e.g., Open Science Collaboration, 2015).

Let me briefly discuss the case of replication as an example of how it is (im)possible to reflect on scientific practice when the normative obligation of replication is not itself the object of empirical scrutiny. Replication is a particularly good case because a classic STS study by Harry Collins (1985) engages with exactly this problem. Accordingly, a comparison can illustrate the difference between reflecting on scientific practice with standard principles of quantitative psychology and doing so with STS. Philosophically it seems rational and sound to require that a study be replicable before you accept its results. However, based on his empirical studies of work in a physics laboratory, Collins showed in the early years of STS that even in physics, replication is anything but straightforward. For one laboratory to replicate the experiment of another, not only must the equipment be the same in both experiments, but both must also utilise similar skills, shared vocabulary, identical material infrastructures and even training with the very same experimental apparatus. The reasons for non-replicability are often found in material infrastructures and practical operations that are not considered part of the scientific apparatus. More than questioning the relevance of replication, Collins' laboratory study (along with many others) indicates that what seems logically and rationally right or wrong about scientific conduct from a philosophical (and scientific) point of view, may in practice turn out to be irrelevant and governed by processes other than those expected.

Based on Collins's and other so-called *laboratory studies*, STS have highlighted the weakness of quantitative science to reflect on scientific practice, when taking their point of departure in normative principles for scientific conduct. From an STS perspective it is sad, therefore, to observe how psychological science currently seeks to enhance and reinforce such principles, e.g., for replication, without founding this on any empirical evidence for the principles' actual effectiveness in scientific practice. The faith in standard normative principles relies, as discussed above, on the regional constitution of quantitative knowledge as the proper procedure for explaining other forms of knowledge while not itself being in need of explanation. When the founding principles for scientific conduct are excluded from reflection, it is only then possible for psychological scientists to seek other procedures for complying with the already set principles, not to test the principles through empirical study of their actual functioning in practice.

## Key STS Concepts for Reflecting on Scientific Practice

What does STS have to offer that will enable psychology to reflect on its own practices and to rearrange its regional constellation with the world? In this section, I present three core concepts that have been developed over the past 40 years and which in my view are central for placing the psychological knowledge producer quite differently in the world from the regional constellation of quantitative psychology.

## Symmetry

The first concept I will discuss was suggested by David Bloor in 1976.[1] Bloor argued against philosophers, historians and sociologists of science who tend to explain true knowledge according to other principles than false knowledge. A contemporary example would be creationist ideas claiming that human life originates from acts of divine creation, as opposed to evolutionary biology's assertion that homo sapiens have developed through natural selection and evolution. The scientific community (along with the vast majority of people educated in Western thought) rejects creationism as a notion rooted in religious beliefs considered to be founded on social and cultural thoughts, contrary to the evolutionary biology argument which is founded on empirical and scientific facts. Following Bloor's symmetry principle, we need to reject this argument, regardless of which side of the creationist-evolutionist debate we support. On the one hand, the idea behind the argument is that as long as someone is being reasonable and following scientific principles, then this in itself guarantees the best explanation. As if we were "rational animals and we naturally reason justly and cleave to the truth when it comes within our view. Beliefs that are true then clearly require no special comment. For them, their truth is all the explanation that is needed of why they are believed" (Bloor, 1991, p. 11). On the other hand, this conceptualisation implies that the only reason not to follow these principles are social, cultural and psychological factors and indeed errors. Put differently, this scheme says that because the methods and logic applied by evolutionary biology are rational, their conclusions are necessarily true. In parallel, because the methods and logic applied by creationists are irrational, their conclusions are necessarily false. The problem in this line of argument is obviously that exactly the same argument can be put forward from the creationist side: their methods and logic are rational, and thus their conclusions are true; while the methods and logic of evolutionary biology are affected by scientific

[1] I here refer to the 1991 2nd edition.

culture, hegemonic social structures and scientists with their sights set on fame and status, and thus their conclusions are irrational. This scheme, Bloor argues, is false. If we accept that true and false knowledge are to be explained according to different factors (rationality explains true knowledge; social, cultural and psychological factors explain false knowledge) then we also need to state what is true and false before investigating truth and untruth. Or put differently: this scheme defines the result of the investigation prior to the investigation.

Instead Bloor (1991) argues that not only false and irrational knowledge, but indeed also true knowledge and scientific facts, are all influenced by social, cultural, psychological and rational factors, and that in order to explain true knowledge we need to refer not only to rationality and logic, but also to social, cultural and psychological dynamics. What social structures and interaction, which cultural habits and imaginaries, and what psychological dynamics convince people that evolutionary biology as a matter of course is true? How does this understanding spread across distance and come to be maintained over time? Indeed, we need to explain true knowledge, Bloor emphasises. As expressed in the quote above, Bloor states that we tend not to see any reason for explaining truth and rationality; only the irrational or erroneous needs explanation. Speaking metaphorically, he makes the ironic statement that, "when a train goes off the rails, a cause for the accident can surely be found. But we neither have, nor need, commissions of enquiry into why accidents do not happen" (Bloor, 1991, p. 8). However, Bloor's point is that we *do* need explanations for why accidents do not happen. We need to understand how it is possible for rational knowledge to maintain its status as true, how it can hold strong as exceptional and superior over the many other modes of knowing. It is not a valid argument to say that a statement is accepted and gains status because it is true. Rather, Bloor argues, a statement is considered true because it is accepted and has gained status.

Bloor's argumentation resonates with my discussion above, about the principles for quantitative psychological knowledge production allegedly needing no explanation in contrast to the knowledge production that follows other principles and practices for knowing and understanding the world. Bloor's *symmetry principle* insist on applying the same vocabulary and criteria for explaining false and true knowledge, and that no form of knowledge production should be immune to the need for explanation. His critique questions the regional constellation of quantitative psychological knowledge production as principally different and separate from Other forms of knowledge production. The symmetry principle places psychological scientific knowledge production on the same epistemic footing as social, cultural and mundane everyday psychological reasoning.

## General Symmetry

While Bloor's argument was based less on empirical research, Latour (Latour & Woolgar, 1986) discovered through ethnography in a neuroendocrinologist's laboratory that not only is scientific knowledge production strongly influenced by social interactions and cultural values, it is also profoundly shaped by material apparatus. Bruno Latour and Steve Woolgar discovered that scientists never observed their object of study "in itself"; rather, the substances they were studying were always put into apparatus that were connected to *inscription devices,* which showed or printed out numbers, colours, graphs, etc. Scientists would examine these inscriptions in accordance with the standards and scales of the inscription devices. Based on the assessments, they would make other apparatus process the inscriptions further, which would result in new inscriptions that scientists would assess, etc. Latour and Woolgar thus concluded that scientific knowledge production is only to a very limited degree a mental activity; much more is it a material and technical activity. This conclusion is not in opposition to understanding scientific knowledge production as social and cultural, as Bloor (1991) did. On the one hand, Latour and Woolgar also observed that social and cultural activities contribute to knowledge production. On the other hand, they note that scientific apparatus is not only technical, but also social. The designs and scales of laboratory apparatus are developed over time and across laboratories, disciplines and enterprises, as a result also of social, economic and cultural dynamics. Therefore, scientific knowledge production is, in Latour and Woolgar's vocabulary, a *socio-material* endeavour.

Based on these and other observations, Latour (1987) came to coin the notion of *general symmetry*. While he supports Bloor's principle about using the same vocabulary to account for both true and false and for both scientific and non-scientific knowledge, he notes that Bloor neglects the material aspects of knowledge production. Not only is apparatus involved in shaping knowledge, but so is also the object of study (which in the natural sciences is often material). Scientists do not shape their apparatus and scales simply out of social and cultural conventions and negotiations, Latour argues. Rather, they also have to negotiate with their object of study and adapt their vocabularies, their ways of observing, their priorities and their social interaction to the type and extent of engagement permitted by this object.

Transferring Latour's line of thinking to psychological science requires a twist, since we would not normally understand psychology's object of study as material. Nonetheless, following Latour's logic the psychological phenomenon we study as psychological scientists becomes a contributor to our scientific knowledge on the same level as our vocabulary, our apparatus, our methods, our social interactions and our cultural imaginaries. Latour (e.g., 1993) proposes an imaginary of *socio-material networks* in which phenomena of all kinds (material, non-material, technical, speech, social interactions, cultural values, etc.) are interrelated. It replaces the idea that our knowledge of a phenomenon is a representation of that phenomenon and thus on a different level of reality than the phenomenon represented by the idea that phenomena of all kinds are interrelated and mutually define each other. Accordingly, when we invite a subject into a laboratory and conduct an experiment based on which we write a report, we would in Latour's view understand these material, social and intellectual activities as extensions and modifications of the socio-material network that makes up our laboratory and our report, including all the interactions, discussions, vocabularies, apparatus, etc. that contribute to producing these. The experimental subject and the practices and networks in which this person is involved are also extended and modified. In other words, Latour envisions our research report—and thus our scientific knowledge—as an extension (rather than a representation) of socio-material networks making up both scientific and non-scientific domains.

Through the socio-material extensions that make up their scientific knowledge, psychological (and other) scientists intervene in and add to the constitution of (admittedly a tiny part) of the world.

Returning to the constellation of scientific psychology and the phenomena it studies, we realise that if we follow this way of thinking, much more than the outcome of our knowledge production is at stake. Among other things, the research design, our vocabulary and our methods contribute to constituting the psychological phenomena. They are anything but neutral techniques that provide an image of the phenomenon "in itself". Vinciane Despret (2004) notes that each research design *makes* phenomena *available* in specific ways. How do our research designs enable psychological phenomena to present themselves? Do our apparatus provide the psychological phenomena with the best possible resources to show us what they are? Do they enable the phenomena to surprise us and present themselves differently from what we expect? Have we thoroughly investigated through which socio-material arrangement they would unfold most effectively? The concept of socio-material networks urges us to ask these kinds of questions which imply an intricate entanglement of the research design with the phenomenon studied (cf. Huniche & Sørensen, 2019b). With this second term (*general symmetry*) we have not only removed the boundary between the psychological scientists and the social beings we study, and thus placed psychological knowledge production on the same level as other types of knowledge production, we have even come to understand our methods, theories and material scientific equipment as *configured together* (Suchman, 2007) or *intra-acting* (Barad, 2007) with the thinking, feeling and acting of our research subjects. We have come to understand psychological scientific practice as intimately entangled with the world of phenomena it studies, dependent on them and responding to them. In this scheme of thoughts, investigating phenomena of the world implies defining one's own position of observation, deeply influenced as it is by the research phenomena. On the one hand, self-reflection thereby becomes an inherent part of scientific practice, and on the other psychological science becomes deeply responsive to the world around it, capable of engaging with its evolutions and surprises.

## Multiplicity

Roar Høstaker (2005) writes that "Latour's solution is to introduce a principle of a general symmetry by which both objects/nature and society are explained simultaneously. When a new scientific fact enters the world, not only has nature changed, but also society and the social actors" (p. 6). As noted above, the notions of object or nature in this quote apply to psychological phenomena as well. The principle of general symmetry implies that psychological phenomena, (similar to nature and other objects) and the knowledge about them are socio-materially produced through the specific network they are (or have become) part of, for example through our research designs. This also implies that before and after becoming entangled with psychological knowledge production, psychological phenomena *differ* from how they are when entangled with the specific socio-material network of the psychological investigation.

Most notably Annemarie Mol (2002) has made clear how both knowledge and phenomena—arteriosclerosis in her study—thus come to have *multiple* ontologies. Based on her hospital ethnography, Mol describes how the disease was something different—both in discursive, social, cultural and material terms—in the clinician's consultation room from the arteriosclerosis observed under the microscope of the pathologist. And it was different again in other socio-material constellations of the hospital. Still, she emphasises that the different enactments and ontologies of the disease are related, and continue from and extend each other, and accordingly arteriosclerosis remains *one* phenomenon (cf. Schank & Sørensen, 2017). Applied to psychological science, this means that we would need to think of subjectivity, identity, aggression, empathy or other psychological phenomena we study as being co-constituted with our knowledge about the phenomena, our identity as researchers, etc. However, these are constituted not only socio-materially—as indicated by the principle of general symmetry—but also as emerging differently—but relatedly—over time and place. The continuous assembling and reassembling of the socio-material networks that make up psychological phenomena also continuously recompose and redefine subjectivities, materialities, social relations and cultural values together with the reassembling of the psychological phenomena. This line of thinking gives up on the dream of universal insights and replaces these with socio-materially contingent insights.

This third and last STS term (*multiplicity*), proposed as a means of enabling psychology to reinvent its constellation with the world it studies, adds to Bloor's placing of psychological knowledge on the same footing as the knowledge of the people studied, and to Latour's intimate entangling of psychological knowledge with the world of phenomena it studies. Mol's concept of multiplicity of phenomena and of psychological knowledge enables the discipline to handle the compound and changing interrelations of which any phenomenon—including any psychological knowledge—is a part. John Law (2004) calls it the "mess of social science", which is at the same time the mess of the world. While psychological science—like any science—will necessarily have to refine and cut (Strathern, 1996) the phenomena we study in order to contribute a new and relevant understanding of them, Mol teaches us that there is no need to argue that this is the only way of approaching the phenomenon, or even the best. The task will be to indicate the specificity of the provided understanding, its ability to interrelate with other understandings and practices, and its contribution to the world of phenomena in question.

## Psychology in the World

I opened this chapter by outlining how large parts of quantitative psychology seem to stick to a modern organising principle although society is shifting away from modern rationales. I described how quantitative psychological knowledge production establishes a constellation of psychology separate and different from Other modes of knowing, which are thus in need of explanation. I argued that this regional constellation inhibits psychology's ability to deal with difference and alterity, while this ability is centrally needed today. I continued by indicating how vocabularies of standardised quantitative psychology obstruct the discipline from starting to question its own scientific practices. I proposed three STS concepts—symmetry, general symmetry and multiplicity—which I argued might help psychology to better reflect on, and modify, its own relationship with the phenomena it investigates; and to stay tuned to today's world.

Quantitative psychology has played a central role in this chapter, mainly as an object of critique. Critique of quantitative psychology is widespread. For this reason, it is important to emphasise that not only do I have great respect for quantitative psychology, but I go so far as to believe that many of its scientific principles are essential for the discipline. My argument is not meant to undermine quantitative psychology. Rather, I suggest a move away from proposing a distanced critique of quantitative psychology—which itself would produce a regional relationship to this area of psychology—to attempting *critical proximity* (Birkbak, Petersen, & Jensen, 2015). This term indicates an approach to a phenomenon—in this case quantitative psychology—that one might at first find problematic. By hesitating instead of rejecting the phenomenon, the researcher becomes able to engage with the *concerns* (Latour, 2004) one shares with it—in this case the study of psychological phenomena. The critical proximity involved in this engagement may generate suggestions to alter core points of the phenomenon, and accordingly get closer to it and create opportunities for co-existence. Engaging with quantitative psychology through three notions of STS suggests a continuation of this area of psychology while revising aspects that have become untimely. An example is the mono-lingual insistence on quantitative vocabulary. There is nothing wrong in principle with quantifications of psychological subject matters, as some colleagues tend to argue (e.g., Markard, 2017). As stated above, it is impossible to make an analysis without modifying the subject matter analysed and without focusing on specific aspects and neglecting others. Quantification does so, which is entirely acceptable; indeed, it is necessary. Accordingly, the STS approach suggests to recognise that relevant psychological insights can be achieved in several ways. There is no compelling reason why quantification should be the most celebrated way of understanding psychological subject matters, but neither is there any good reason to avoid quantification. If we add to quantification the principle of accounting for the specificity of this way of coming to know a subject matter—i.e. account for the consequences that adding quantification to a subject matter has for the way in which it can possibly relate to other phenomena (including to ourselves as scientists)—then we can bring quantitative psychology completely into alignment with the STS vocabulary I have proposed in this chapter. A core advantage of this vocabulary is that it enables psychology as a discipline to remain attentive to the way in which it relates to the phenomena it studies, and it enables it to adapt its methods to these phenomena, and thus to remain timely. In order to reinvent psychological science so that it meets the challenges of today's world, the discipline needs to understand itself as part of the world it inhabits.

## References


Asplund, J. (1985). *Tid, rum, individ och kollektiv* [Time, space, individual and collective]. Stockholm, Sweden: Liber.

Baker, M. (2015, August 3). Over half of psychology studies fail reproducibility test. *Nature*. https://doi.org/10.1038/nature.2015.18248.

Barad, K. (2007). *Meeting the universe halfway: Quantum physics and the entanglement of matter and meaning*. Durham, NC: Duke University Press.

Barry, A., & Born, G. (2013). Interdisciplinarity: Reconfigurations of the social and natural sciences. In A. Barry & G. Born (Eds.), *Interdisciplinarity: Reconfigurations of the social and natural sciences* (pp. 1–56). London, UK: Routledge.

Beck, U. (1992). *Risk society: Towards a new modernity*. London, UK: Sage.

Beck, U. (2016). *The metamorphosis of the world: How climate change is transforming our concept of the world*. Cambridge, UK: Polity Press.

Beck, S., Niewöhner, J., & Sørensen, E. (Eds.). (2012). *Science and technology studies: Eine sozialanthropologische Einführung* [Science and technology studies: A social-anthropological introduction]. Bielefeld, Germany: Transcript.

Birkbak, A., Petersen, M. K., & Jensen, T. E. (2015). Critical proximity as a methodological move in techno-anthropology. *Techné: Research in Philosophy of Technology, 19*(2), 266–290. https://doi.org/10.5840/techne201591138.

Bloor, D. (1991). *Knowledge and social imagery*. Chicago, IL: University of Chicago Press.

Brown, N. J., & Heathers, J. A. (2016). The GRIM test: A simple technique detects numerous anomalies in the reporting of results in psychology. *Social Psychological and Personality Science, 8*(4), 363–369. https://doi.org/10.1177/1948550616673876.

Colin, F. C., Dreber, A., Holzmeister, F., Ho, T.-H., Huber, H., Johannesson, M., ... Wu, H. (2018). Evaluating the replicability of social science experiments in nature and science between 2010 and 2015. *Nature Human Behaviour, 2*, 637–644.

Collins, H. (1985). *Changing order: Replication and induction in scientific practice*. London, UK: Sage.

Deleuze, G. (1979). Introduction. In J. Donzelot (Ed.), *The policing of families: Welfare versus the state*. London, UK: Hutchinson.

Derksen, M. (2019). Putting Popper to work. *Theory & Psychology*. Online first. https://doi.org/10.1177/0959354319838343.

Despret, V. (2004). The body we care for: Figures of anthropo-zoo-genesis. *Body & Society, 10*(2–3), 111–134.

Etz, A., & Vandekerckhove, J. (2016). A Bayesian perspective on the reproducibility project: Psychology. *PLoS One, 11*(2). https://doi.org/10.1371/journal.pone.0149794.

Hess, D. J. (1997). *Science studies: An advanced introduction*. New York, NY: New York University.

Høstaker, R. (2005). Latour—Semiotics and science studies. *Science Studies, 18*(2), 5–25.

Huniche, L., & Sørensen, E. (2019a). Psychology's epistemic projects. *Theory & Psychology, 29*(4), 441–448. https://doi.org/10.1177/0959354319863496.

Huniche, L., & Sørensen, E. (2019b). Phenomenon-driven research and systematic research assembling: Methodological conceptualisations for psychology's epistemic projects. *Theory & Psychology, 29*(4), 539–558. https://doi.org/10.1177/0959354319862048.

Jensen, T., & Sørensen, E. (1995). *Fra tid til anden* [From time to time]. Retrieved from https://www.sowi.rub.de/mam/content/cupak/jensen_sorensen_1995_fratidtilanden.pdf.

Latour, B. (1987). *Science in action: How to follow scientists and engineers through society*. Cambridge, MA: Harvard University Press.

Latour, B. (1993). *We have never been modern*. Cambridge, MA: Harvard University Press.

Latour, B. (2004). Why has critique run out of steam? From matters of fact to matters of concern. *Critical Inquiry, 30*(2), 225–248.

Latour, B., & Woolgar, S. (1986). *Laboratory life: The construction of scientific facts* (1st Princeton paperback print). Princeton, NJ: Princeton University Press.

Law, J. (2004). *After method: Mess in social science research. International library of sociology*. London: Routledge.

Markard, M. (2017). Standpunkt des Subjekts und Gesellschaftskritik. Zur Perspektive subjektwissenschaftlicher Forschung [Subject standpoint and social critique. A subject-oriented perspective]. In D. Heseler, R. Iltzsche, O. Rojon, J. Rüppel, & T. D. Uhlig (Eds.), *Perspektiven kritischer Psychologie und qualitativer Forschung* (pp. 227–244). Wiesbaden: Springer.

Mol, A. (2002). *The body multiple: Ontology in medical practice: Science and cultural theory*. Durham, NC: Duke University Press.

Mulkay, M., & Gilbert, G. N. (1981). Putting philosophy to work: Karl Popper's influence on scientific practice. *Philosophy of the Social Sciences, 11*(3), 389–407. https://doi.org/10.1177/004839318101100306.

Open Science Collaboration. (2015). Estimating the reproducibility of psychological science. *Science, 349*(6251), 943. https://doi.org/10.1126/science.aac4716.

Pashler, H., & Wagenmakers, E.-J. (2012). Editors' introduction to the special section on replicability in psychological science: A crisis of confidence? *Perspectives on Psychological Science, 7*(6), 528–530. https://doi.org/10.1177/1745691612465253.

Porter, T. M. (1995). *Trust in numbers: The pursuit of objectivity in science and public life*. Princeton, NJ: Princeton University Press.

Rose, N. S. (1999). *Governing the soul: The shaping of the private self*. London, UK: Free Association Books.

Schank, J., & Sørensen, E. (2017). Praxeographie – Einführung [Praxeography—An introduction]. In S. Bauer, T. Heinemann, & T. Lemke (Eds.), *Science and technology studies. Klassische Positionen und aktuelle Perspektiven* (pp. 407–428). Berlin: Suhrkamp.

Schön, D. A. (1983). *The reflective practitioner: How professionals think in action*. New York, NY: Basic Books.

Sismondo, S. (2004). *An introduction to science and technology studies*. Malden, MA: Blackwell.

Sørensen, E. (2007). The time of materiality. *Forum Qualitative Social Research, 8*(1), Art. 2. https://doi.org/10.17169/fqs-8.1.207.

Sørensen, E. (2008). *The materiality of learning: Technology and knowledge in educational practice*. Cambridge: Cambridge University Press.

Strathern, M. (1996). Cutting the network. *The Journal of the Royal Anthropological Institute, 2*(3), 517. https://doi.org/10.2307/3034901.

Stroebe, W., Postmes, T., & Spears, R. (2012). Scientific misconduct and the myth of self-correction in science. *Perspectives on Psychological Science: A Journal of the Association for Psychological Science, 7*(6), 670–688. https://doi.org/10.1177/1745691612460687.

Suchman, L. (2007). *Human-machine reconfigurations: Plans and situated actions*. Cambridge: Cambridge University Press.

Verran, H. (1999). Staying true to laughter in Nigerian classrooms. In J. Law & J. Hassard (Eds.), *Actor-network theory and after* (pp. 136–155). Oxford, UK: Blackwell.

Verran, H. (2001). *Science and an African logic*. Chicago, IL: University of Chicago Press.

# 10

# The Social Production of Evidence in Psychology: A Case Study of the APA Task Force on Evidence-Based Practice

Nathalie Lovasz and Joshua W. Clegg

In recent decades, the concept of evidence-based practice (EBP) has come to occupy a central role in the discourses surrounding psychological treatment. Insurance companies, for example, increasingly require proof that clinicians are engaging in EBP (Levant, 2005). And both the American Psychological Association and the Canadian Psychological Association's accreditation guidelines for training programs in clinical psychology require that students be instructed in the use of EBP (APA, 2006b, p. 7; CPA, 2011, p. 21). The concept of EBP has thus come to occupy a gatekeeping role and carries financial and practical implications for practitioners, researchers and trainees.

The concepts of evidence and evidence-based practice, however, differ from other forms of gatekeeping in clinical psychology

N. Lovasz (✉)
Toronto DBT Centre, Toronto, ON, Canada

J. W. Clegg
John Jay College and the Graduate Center, CUNY, New York, NY, USA
e-mail: jclegg@jjay.cuny.edu



K. C. O'Doherty et al. (eds.), *Psychological Studies of Science and Technology*, Palgrave Studies in the Theory and History of Psychology,
https://doi.org/10.1007/978-3-030-25308-0_10

(e.g., academic credentials, board certifications) in that they are much less clearly defined and universally agreed upon. When a clinician claims to be employing EBP it is not always immediately apparent what exactly this means.[1] There is no formally defined process for demonstrating the legitimacy of such a claim (as in the passing of exams and evaluations for academic credentials or certifications).

Some attempts have been made to more formally legislate the meaning and use of evidence in clinical practice, including the division 12 list of Empirically Supported Treatments (ESTs) and policies on EBP developed by both the American and Canadian Psychological Associations. Probably the most commonly agreed upon and systematic articulation of EBP in psychology came out of the proceedings of a Presidential Task Force on Evidence-based Practice, convened by the American Psychological Association (APA) in 2005. The policy document generated by the Task Force approximates a formally defined set of standards for the use of evidence and for EBP in clinical practice. As such, the proceedings and products of this Task Force represent a unique concretion of practices and discourses whose function has been to stabilize a particular way of defining and enacting "evidence".

We focus in this chapter on the history of the Task Force as a way of tracing the development and deployment of a particular institutionalization of "evidence". Our argument is that the manner in which evidence and EBP came to be defined by the APA Task Force and the way EBP policy documents have since been applied and utilized is best understood as a continually unfolding sociopolitical process that has contributed to the social production of "evidence" as a concept (and practice) in psychology. In the sections that follow, we outline the political, practical and interpersonal considerations that shaped the proceedings of the APA Task Force on EBP and the policy document that emerged from those proceedings. We also explore the social forces that shaped how the Task Force policy has been interpreted, applied and utilized by the psychological community in the time since its publication, which in turn

[1] To be fair, what scientists or philosophers of science mean by "evidence" is not clear nor universally agreed upon either.

have shaped the ways that EBP and the concept of evidence are understood by the psychological community at large.

## Methods

The analysis in the current chapter is mainly based on archival and qualitative research conducted by the first author. We reviewed working documents, reports, and minutes from the proceedings of the Presidential Task Force on EBP obtained directly from the APA. We further contextualized and interpreted information gathered during the archival review with information gained during interviews with five of the members of the Task Force. Members who agreed to participate (all members were invited to participate) were interviewed by phone or in person between February and August of 2011. The interview guide consisted of a series of open-ended questions emailed to participants for their review to provide participants a chance to alter or remove questions from the interview guide as they saw fit; no participants chose to exercise this right. Interviews were semi-structured, audio-recorded and transcribed verbatim for content (non-verbal responses were included only insofar as they enhanced the understanding of content). As a part of developing the current analysis, we reviewed and analyzed interviews and other materials for relevant content. As is always the case in qualitative research or historical analysis, this chapter presents a narrative based on our interpretations of documents and interview transcripts. We do not strive to provide an epistemically neutral nor objective account. We do not make absolute claims about the history of EBP but rather present our perspectives, interpretations, and theoretical understandings of the documents reviewed.

## Political Considerations That Prompted the Formation of the Task Force

The APA Presidential Task Force on EBP was formed in response to a number of social and political pressures within and outside of the discipline of psychology. Outside of psychology, EBP as a concept originated in medicine (Sackett, 2000) in an attempt to address concerns that scientific research examining and comparing medical interventions was insufficiently employed as a basis for medical decision-making. The concept of evidence-based medicine fell on fertile ground amidst growing concerns about the rising costs of health care as well as inconsistencies in health care delivery.

In 2001, the Institute of Medicine (IOM), a non-profit health care policy advisory arm of the National Academy of Sciences published a report by its Committee on the Quality of Health Care in America. The report described widespread disparities in medical training and health care delivery and suggested evidence-based medicine as a training and practice standard to remedy these issues. The committee adopted the following definition of evidence-based medicine from Sackett and colleagues (2000): "Evidence-based practice is the integration of best research evidence with clinical expertise and patient values" (p. 147). This definition was to become highly influential in the deliberations on EBP in psychology (see below).

By the turn of the millennium, evidence-based medicine had become a catch-phrase, although its central premises, feasibility, and practical applicability were also beginning to come under attack (e.g., Eddy, 2005). Despite these criticisms, EBP was becoming increasingly important in health care policy and so various other health care disciplines began to develop their own guidelines for EBP. These developments concerned psychologists who foresaw the need to legitimize psychological treatments within the same framework.

Psychologists had long debated the relative importance of objective data versus subjective inference in psychological assessment and clinical decision-making.[2] For most of psychology's history, some psychologists have favored actuarial approaches to decision-making while others defended the importance of clinical judgment (for a more detailed discussion of this see Meehl, 1954). Thus, in parallel to the development of evidence-based

[2]A discussion of these broader debates in Psychology is beyond the scope of the current paper but the reader is pointed to debates between actuarial vs. clinical judgment based assessment, psychoanalysis vs. behaviorism, idiographic vs nomothetic approaches to assessment and treatment and projective vs. objective assessment methods.

medicine, some psychologists had already attempted to establish an evidence-base for psychological practices through the development of treatment guidelines and the "empirically validated treatment" movement.

The empirically validated treatment movement originated in a task force on the promotion and dissemination of EST, formed by Division 12 (the Society of Clinical Psychology) of the American Psychological Association during the presidency of David Barlow in 1993. In Barlow's (2011) words:

> The purpose of that task force was to really make all the mental health professions and the public at large and policy makers more aware that we have very strong treatments for a variety of disorders that perhaps weren't widely recognized and weren't being widely administered. (personal communication, August 6, 2011)

EST Task force members generated a stipulative definition (Chambless et al., 1996) that included two types of empirically validated treatments—well-established treatments and probably efficacious treatments—and defined criteria that would establish treatments in each category (e.g., support from two RCTs). These criteria privileged certain kinds of methods (RCTs codified as the "gold standard"), and so institutionalized a methodologically hierarchical approach to evidence.[3]

One of the most controversial components of the task force's work was a list of psychological treatments for specific psychological problems that fit the criteria for well-established and probably efficacious treatments. The rationale for this list, according to Chambless et al., (1996), was the sense that most clinicians did not have the time to locate, review, and consider relevant research findings. As such, providing them with lists of treatments supported by research was considered a valuable shortcut facilitating evidence-oriented practice. The list was preliminary, based on task force members' knowledge, and was not intended to be

[3]Throughout this chapter, we contrast a methodologically pluralistic approach to evidence—that is, an approach where different methods and the different kinds of evidence these produce are understood to be complementary and parallel—with a methodologically hierarchical approach to evidence—that is, an approach where certain methods and the evidence they produce (principally RCTs) are considered superior to other kinds.

exhaustive. In fact, the task force recommended that a more exhaustive list be generated and updated regularly.

The intended audience, which included the general public, practitioners, and third-party payers, however, did not apply the list in that spirit. Instead, they took a dichotomous view of the effectiveness of treatments (i.e., effective or ineffective). This approach was typified by managed care and health maintenance organizations who began to use the list as a basis for decisions about treatment funding, thus limiting what treatments could be reimbursed (Barlow, personal communication, 2011). Attempts by Division 12 to ameliorate these effects, including emphasizing the tentative nature of their list and abandoning the "empirically validated treatments" terminology (in favor of EST), did little to change such practices.

Growing concerns about the social and disciplinary implications of EST set the stage for the formation of the broader APA Task Force on Evidence-based Practice. Misuses of the division 12 EST lists left many suspicious of EBP in general. As David Barlow, framed it:

> There were some on the Task Force who, quite reasonably, felt that unless they employed a very small narrow range of procedures they might not, that practitioners might not be reimbursed for their practice; because there had been some abuses by managed care companies leading up to this point. The managed care companies basically said you can only do so many sessions and it has to be something on our list, anything else is not reimbursable so it was that kind of fear. (personal communication, August 6, 2011)

Other Task Force members voiced similar concerns. According to Steven Hollon, for example, "there was concern that livelihoods were moving out of their control as opposed to the society trusting a licensed clinical psychologist" (S. Hollon, personal communication, April 26, 2011). These fears, coupled with the rising cross-disciplinary influence of EBP, suggested to many that the time had come for a more formal approach to EBP in psychology.

The 2005 formation of the APA Task Force on Evidence-based Practice constituted the APA's response to these developments and marked the formal entry of "Evidence-based practice" into the discourse surrounding, and policies governing, psychological practice. Ronald Levant, the 2005 president of APA, was a key figure in the formation of this Task Force and his motives also reflected concerns over EST:

> I formulated the idea that we needed to develop a policy analogous to that of the IOM policy, which we would call evidence-based practice in psychology, so that's kind of how it originated. I felt that the Division 12 lists were potentially harming psychology, although I understand that Dave Barlow who is a friend and a colleague did not have that intention. In fact, his intention was very clearly in setting up that task force, was to counteract the, at that time, 1995/1994 the overwhelming trend towards viewing medications as far more effective than psychological treatments and the treatment of mental illness. So, that was his intent, and that's a valid one and one I support, but I felt that it was having the unintended impact of forcing psychologists into procrustean beds that really didn't fit effective psychological practice, so that's really how it got started. (personal communication, June 27, 2011)

The EBP movement in psychology therefore emerged in response to pressures to establish the scientific legitimacy of psychological treatments as well as to concerns that criteria used to establish that legitimacy could be abused if they were too narrow or otherwise misunderstood. Understood in this broader context, the Task Force can be seen as an attempt to establish and secure the boundaries of psychological practice—to protect disciplinary autonomy, individual livelihoods, and existing centers of institutional influence (e.g., the APA) against incursions from elements within psychology (e.g., the EST movement) and without (e.g., insurance companies).

## Practical and Institutional Factors That Impacted the Proceedings of the Task Force

Given these political pressures and the high stakes involved, it is not surprising that a sense of urgency fueled the proceedings of the EBP Task Force. The need to act pre-emptively created time pressures on

Task Force members and this led, in part, to a decision to work from the IOM definition of EBP. Task Force leaders also responded to fears of overly narrow definitions of evidence by explicitly tasking the group with broadening the range of evidence considered by policymakers. Together, these considerations instituted a priori constraints on how the Task Force could pursue their deliberations.

Task Force deliberations occurred relatively quickly over the course of approximately one year. Task Force members met face-to-face on two occasions in October 2004 and January 2005 (Presidential Task Force on Evidence-based Practice, Spring Consolidated Meeting Agenda Item, unpublished document, March 2005). According to APA President at the time, Ronald Levant, at the first meeting members were given a short amount of time to present their perspectives on EBP (they were limited to 3–5 slides to reduce "speechifying") and a chance to debate their respective views. During the remainder of the proceedings, the Task Force broke into small writing groups to work on various parts of the document and reconvened to discuss and review drafts as a whole group (APA Presidential Task Force on Evidence-based Practice Meeting Agenda, unpublished document, January 2005b). Between the meetings of the Task Force, phone meetings and email discussions were used to revise and rework various sections of the Task Force documents.

It is clear, then, that the Task Force sought to quickly generate a policy document and this required focused effort. In addition, and possibly in part because of the time pressures described above, rather than generating a psychological policy from scratch the Task Force was charged with using the widely accepted IOM definition. Despite their diverse backgrounds and perspectives, Task Force members were asked to use that definition as a starting point in their deliberations (starting with the assumption that the definition is "generally acceptable"). Members were also asked to expand on this definition by incorporating greater attention to patient preferences, multicultural perspectives, and a "practitioner scientist" role for clinicians.

A final constraint on Task Force deliberations was the mandate, inspired at least in part by desires for a more methodologically pluralistic approach to evidence, to broaden the range of evidence policymakers consider when making health care decisions. This emphasis is evident in the Draft Policy Statement submitted to the APA Council of Representatives:

> In this report, The Task Force hopes to draw on APA's century-long tradition of attention to the integration of science and practice by creating a document that describes our fundamental commitment to sophisticated evidence-based psychological practice and takes into account the full range of evidence that policy makers must consider. (APA Presidential Task Force on Evidence-based Practice, unpublished document, August 2005a)

At the outset, then, the institutional framework guiding Task Force deliberations was shaped both by practical considerations, like the need for rapid and focused work, and by ideological considerations, like the need to accommodate multiple perspective on evidence.

## Social and Relational Factors That Impacted the Proceedings of the Task Force

The political and ideological pressures already described were enacted not only in the institutional framework for the Task Force, but through the relationships of Task Force members. Members were colleagues who had previously worked on research, projects or committees together and would likely be working together again in the future. Despite considerable member diversity (and disagreement), this collegial dynamic, coupled with pressures to mediate disagreements and reach resolutions, made possible (and necessary) the formulation of a broader policy that could accommodate divergent views.

This broader policy began with a Task Force selection process meant to create an ideologically diverse committee dynamic. Members for the Task Force were selected intentionally for their divergent views and also for their capacity to mediate disagreements and arrive at moderate positions. The main objective in the formation of the Task Force, according to Ronald Levant, was to select members that represented all constituencies active in the caucuses of the APA. This emphasis can again be seen in the Task Force charge:

> The Task Force incorporates scientists and practitioners from a wide range of perspectives and traditions, reflecting the diverse perspectives within the field: Clinical expertise and decision-making; health services research; public health and consumer perspectives; RCT science; full time practice; clinical research and diversity; non-RCT clinical research; health care economics; EBP research/training and applications. (APA, unpublished document, 2004)

In addition to representing all caucuses, Levant described additional considerations: "I wanted people who met two criteria: that they were respected by their caucus or constituency, probably their broader constituency, and that they were statesman-like people who could hear other points of view and could compromise" (personal communication, June 27, 2011). As a result of the criteria described above, Task Force members represented divergent opinions in psychology but were also able to remain conciliatory and moderate in their views.

This does not imply, of course, that Task Force deliberations were without disagreements. Most of the Task Force members interviewed recalled numerous debates, the majority of these focused on the relative importance of research evidence and clinical expertise, on the relative importance of various kinds of evidence in making practice decisions, and on the adequacy of currently existing research evidence in guiding decision-making.

While Task Force members agreed that evidence, clinical expertise, and patient preferences each play an important role in clinical decision-making, the weighting of these three components was highly contested during Task Force deliberations. One of the principal areas of divergence in Task Force members' views concerned the importance of research evidence compared to clinical judgment. In the words of Task Force member David Barlow:

> I think there was a component on the Task Force who clearly adhered to the supremacy of clinical prediction over statistical prediction and there was another group, including myself, who were much more confident in the more empirical statistical prediction, so that was kind of the implicit divergence on the Task Force. (personal communication, August 6, 2011)

Most Task Force members interviewed commented on this divide but agreed that ultimately the Task Force weighted research evidence more heavily than clinical expertise. The Task Force incorporated this weighting into its policy by conceptualizing respect for research evidence as part of clinical expertise, thus making expertise ultimately dependent on a consideration of relevant research evidence. Bruce Wampold described this position:

> So you can't be a clinical expert if you ignore or are ignorant of research evidence. And that was the way their perspective was accommodated, and I think it satisfied the practitioners too. (personal communication, February 22, 2011)

Though these different statements on the relative importance of clinical expertise and research evidence show the accommodations reached by the Task Force, they also reflect differing emphases, differences that were exaggerated in the years following the report.

The second area of debate on the Task Force related to the kinds of evidence to be used in EBP. John Norcross described the key issues:

> I believe I'm an ardent proponent of methodological pluralism, that's largely dependent upon the question being addressed. The entire Task Force embraced that notion, though the devil was lurking in the details. To what extent do we price a controlled outcome study over a naturalistic effectiveness? What are the relative advantages and disadvantages of say systematic case studies? (J. Norcross, personal communication, April 28, 2011)

Task Force members came to different conclusions about these relative degrees of importance or relevance, and this was reflected in differing assessments of currently available evidence:

> I think there was an attempt by some of the Task Force to minimize the evidence we already had with the goal of, to state that you know the kind of evidence we have is very very limited in terms of its applicability to practice and then there were others such as me who said that's not true, the evidence actually is quite good. (D. Barlow, personal communication, August 26, 2011)

In fact, there was no consensus among Task Force members interviewed on the relative value of different kinds of evidence nor whether the list of various types of evidence provided in the Task Force statement was to be interpreted in a hierarchical fashion or, instead, as a list of multiple methods that could be used to generate evidence for different purposes. These differences, though glossed in the Task Force report, were also exaggerated in later years.

Despite the divergences in Task Force members' views on the issues described above, the draft policy statement on EBP includes the following statement:

> Perhaps the central message of this task force report, and one of the most heartening aspects of the process that led to it, is the consensus achieved among a diverse group of scientists, clinicians, and scientist-clinicians from multiple perspectives that EBPP requires an appreciation of the value of multiple sources of scientific evidence. (APA, unpublished document, 2005)

Most of the Task Force members interviewed echoed this sentiment and emphasized the friendly and conciliatory character of Task Force deliberations. Indeed, many reported their discussions and collaborations facilitating the formation of friendship and greater agreement among members. A number of Task Force members also described their own views changing in response to this collegial and pluralistic process:

> A lot of us came to really like and respect the other folks that were sitting in the room from different perspectives. I'd be very surprised if (name of a Task Force member) changed his perspective on the basis of our conversation. I'd be very surprised if I changed my perspective on the basis of our conversation although I gotta say I think I did. I think I have more respect for stuff that I would have dismissed until I worked with those folks closely on the Task Force. (S. Hollon, personal communication, April 26, 2011)

This collegiality and consensus was produced by a deliberative and cooperative process, but also by a deliberately diplomatic approach to disagreement: "the brilliant beauty of the Task Force was that it focused on the gap you know between the two perspectives and the recognition that you know my truth and your truth while very meaningful to each of us respectively are only a small portion of the whole truth" (R. Levant, personal communication, June 27, 2011). Many Task Force members commented on the fact that more philosophical topics, such as debates about the value of specific scientific methods, were avoided by the Task Force in the interest of promoting agreement between members:

> We purposely avoided that [philosophical discussion] because we were gonna quickly get to some irreconcilable differences you know and the different kinds of research evidence, you even see qualitative research there right? Because there were some people on the Task Force that said well qualitative is the only way research should be done, well you know once you take that position you're never going to come to an agreement with people like (names of particular Task Force members). I think so, I think we tried to avoid those things. (B. Wampold, personal communication, February 22, 2011)

All of this diplomacy was ultimately reflected in Task Force documents, which were kept relatively general to allow for broad endorsement. David Barlow described this diplomatic writing process: "Many of the sentences that one person or another would narrate would be too specific to agree on so we had to back off and become more general in how we said it that it'd cover all" (personal communication, August 6, 2011).

Not surprisingly, in their final documents, the Task Force produced a broad and methodologically pluralistic account of EBP. They defined EBP as "the integration of the best available research with clinical expertise in the context of patient characteristics, culture, and preferences" (APA Presidential Task Force on Evidence-based Practice, 2006a, p. 3). Task Force documents endorsed the use of multiple types of evidence, including evidence produced by various research designs. They also provided a list of various types of evidence and the types of questions each type of evidence is suited to address (e.g., clinical observation, qualitative research, systematic case studies, single-case experimental designs, public health and ethnographic research, process-outcome studies, effectiveness studies, randomized controlled trials [RCT's]/efficacy research, and meta-analysis).

The Task Force documents also advocated a more holistic approach to evaluating treatment, emphasizing the importance of clinical expertise and patient characteristics, culture, and preferences. In doing so, the Task Force acknowledged that findings based on groups may not apply to all individuals and that individual characteristics may be relevant to the outcome of interventions. In particular, they conceded the difficulty of applying evidence across populations. They described the multitude of ways in which patients may differ on individual characteristics, culture, and personal preference, and pointed out the importance of considering such factors in EBP.

Clearly, then, the APA statement on EBP provides very general guidelines rather than a set of specific rules. Precisely how to integrate the various components of EBP and their relative importance to psychological practice remain unspecified. In fact, nothing in any version of these statements suggests a hierarchical organization of the three facets of EBP, nor of differing forms of evidence.

## Institutionalization and Implementation in the Years Following the Report

It is not surprising that an intentionally diverse and "statesman-like" process would produce a methodologically pluralistic document, but, as John Norcross said, the devil is in the details. In the years since the Task Force report, the differences minimized in the creation of Task Force statements have become magnified in subsequent attempts to implement those statements. To begin at the end of the story, it is quite clear that, although the implementation of EBP (of any sort) has only just begun, much more has been done to institutionalize a methodologically hierarchical view of evidence than a pluralistic one.

What we are calling the "hierarchical view" is, of course, not monolithic, but reflects an array of positions on the role of evidence in clinical practice. Some of these positions simply assert the superiority of certain kinds of research designs (nearly always RCTs). Others, like the EST movement, legislate the same evidential hierarchy by creating lists of "evidence-based" or "empirically supported" treatments, compiled on the basis of some kind of research design ranking (again, nearly always with RCTs at the top). Despite their different manifestations, in all cases, instances of the "hierarchical view" share a commitment to the superiority of certain kinds of evidence (and the research designs that produce them).

Among the subtlest ways that a methodologically hierarchical view has been institutionalized, are a general set of rhetorical forms that take for granted a hierarchical view of evidence. There is general acknowledgment that the Task Force report (and EBP in general) "is not simply a list of treatments that have demonstrated empirical support" (Beck et al., 2014, p. 413), but is a more global approach to the proper use of evidence in therapeutic practice. Yet, even in the midst of such acknowledgments, authors frequently signal a commitment to a hierarchical view of evidence (with certain therapies, well studied and supported by RCTs at the top of the heap). Bearman, Wadkins, Bailin, and Doctoroff (2015), for example, who provided the caution just quoted, on the same page chided clinical training programs because they "did not require both didactic training and clinical supervision in evidence-based therapies" (p. 413). Similarly, Bearman et al. (2015) lamented (as a sign that EBP has not sufficiently penetrated the discipline) that "Only 20% of PsyD programs ... required clinical supervision in CBT", one of the treatments most widely studied using RCTs (Bearman et al., 2015, p. 14).

Not all rhetorical strategies are so subtle. Lilienfeld, Ritschel, Lynn, Cautin, and Latzman (2013), for example, challenged any pluralistic view of evidence, arguing for the simple superiority of evidence from RCTs and of scientific research relative to clinical experience. The Canadian Psychological Association report on EBP also explicitly endorses an evidential hierarchy (CPA, 2012, p. 6). Other strategies for rhetorically privileging a hierarchical view of evidence include treating the question as resolved and treating a failure to comply with some version of the hierarchical view as the result of misunderstanding, naivete, resistance, or a lack of education—a "science-practice" gap (Dobson, 2016) in need of remediation.

Those who support a more methodologically pluralistic view of evidence have also attempted to "control the narrative", though theirs seems to be the minority position. In their clarification of the Task Force recommendations, Wampold, Goodheart, and Levant (2007) explicitly disavowed a "therapy ranking" approach to evidence, arguing that "terms such as evidence-based treatments are not indigenous to EBPP as defined by the APA" (p. 617). They also argued that the Task Force recommendations "did not dictate the method used to collect data that would form the basis of evidence, nor did it privilege certain types of evidence" (p. 617). Rather, they argued, the report avers that "some methods are better suited for some purposes than for others" (p. 617), and while "clinical trials provide optimal evidence on the particular question of treatment efficacy" (p. 617), they are not necessarily the best kind of evidence for other important questions in psychotherapy research. They also explicitly supported a kind of epistemic pluralism: "practitioners consider both nomothetic and idiographic evidence and need to hold simultaneously scientific and humanistic perspectives, which are 'psychology's dual heritage' (Messer, 2004, p. 586). This is not simply a compromise—it is reflective of the complex task of psychotherapy" (p. 618).

Others have embraced this pluralistic view. Rousseau and Gunia (2016), for example, argued that "in contrast to critiques of EBP as overly valuing randomized controlled trials (RCTs; e.g., Webb 2001), the diversity of possible practice questions necessitates methodological pluralism" (Rousseau & Gunia, 2016, p. 670). Ronald Levant has also continued to champion this view in later years, suggesting that "many different kinds of research designs contribute to EBPP" and that "each of these types of research makes its own unique contribution (APA 2005 Presidential Task Force on Evidence-Based Practice, 2006)" (Levant & Sperry, 2016, p. 20).

These attempts to defend a pluralistic account of evidence notwithstanding, the general narrative seems to bend toward the hierarchical view. As Stewart, Chambless, and Stirman (2018) state, "despite efforts to temper the definition of EBP from skewing too far toward the research evidence (Levant, 2005), many clinicians, researchers, and policymakers appear to consider the construct of evidence-based practice as synonymous with research, evidence-based treatments or empirically supported treatments" (p. 57).

This conflation of EBP and other approaches (like EST) to the use of evidence in clinical practice is likely not simply the result of efforts to institutionalize the hierarchical view of evidence; we suspect that it is also partly a cause of that institutionalization. Because, as Stewart et al. (2018) argued (see above), so many ignore the more holistic and methodologically pluralistic elements in the Task Force definition and, instead, take the simplified hierarchical approach to evidence as roughly identical with EBP, even those claiming an EBP orientation are often enacting something closer to EST.

Beyond the rhetorical strategies discussed so far, the ascendance of the hierarchical view of evidence is likely also due in part to more overt forms of institutionalization. These include integration into: legislation, licensing and accreditation requirements, funding requirements, insurance reimbursement, and evidence-ranking systems. In what follows, we discuss each of these, in turn.

One of the most influential ways that the hierarchical view of evidence has been institutionalized is through professional regulatory requirements. For example, "the APA Standards for Accreditation (American Psychological Association, 2006) include the guideline that students receive training in ESTs, and a Division 12 Task Force report (1995) recommended that training in ESTs should be a requirement for the APA accreditation of training programs" (Thomason, 2010, pp. 30–31). The Council of University Directors of Clinical Psychology Programs also argued "that clinical psychology doctoral programs should provide training in empirically supported treatments" (Forman, Gaudiano, & Herbert, 2016, p. 162).

Another way that the hierarchical view of evidence has begun to be institutionalized is through integration into reimbursement requirements. According to McGrew, Ruble, and Smith (2016), for example, "advocacy efforts to expand insurance coverage of ASD treatment in the United States" have emphasized the "provision of a limited list of EBPs" (p. 248). Levant and Hassan (2008) take this approach to be a general trend, noting that "the American Psychological Association Division of Clinical Psychology (1995; see also Chambless et al., 1998) lists of empirically supported treatments have been referenced by a number of local, state, and federal funding agencies, who are beginning to restrict reimbursement to these treatments, as are some managed care and insurance companies" (p. 658). La Roche and Christopher (2009) have argued that this shift toward requiring EST therapy was precipitated by MCOs who "encouraged and increasingly required" ESTs and thus set a precedent that "many states followed … mandating the use of mental health treatments considered to be evidenced based within Medicaid programs" (p. 397). Thomason (2010) predicts that malpractice insurance providers will soon succumb to this trend and that "third-party payers will eventually reimburse only for empirically supported treatments" (p. 31).

The hierarchical notion of evidence undergirding the EST movement has also been partly institutionalized through integration into funding agency policies. Foster (2015), for example, claims that service delivery managers "increasingly are choosing those therapies which have a greater evidence base; empirically supported treatments" (p. 34). Federal and foundation grant funding is also often arbitrated on the basis of evidentiary hierarchies (La Roche & Christopher, 2009).

Taken together, these forms of institutionalization solidify a certain view of evidence; even more, they enact or plot that view of evidence within the political and material networks that arbitrate mental health treatment. They accrete into a taken-for-granted and self-perpetuating ecosystem. Thus, even the systems for assessing the "quality" of evidence have come to institutionalize the hierarchical view. In particular, the relatively recent standardized evidence assessment protocols like GRADE (Guyat et al., 2008) formalize evidence hierarchies (GRADE, for example, considers only meta-analyses and so a priori excludes single-case designs; Forman et al., 2016), and so "close the loop" against a more pluralistic view.

Of course, even though the hierarchical view of evidence has been institutionalized in various ways, it is not clear how much these developments are actually constraining practice. In fact, Stewart et al. (2018) argue that "a small but significant literature from multiple samples has accumulated indicating that clinicians discount research evidence in favor of using clinical experience to inform treatment decisions" (p. 57). Putting the case more starkly, Forman et al. (2016) claim that "most clinicians still do not utilize ESTs in their practice" (p. 162). Among those concerned with EBP, the standard way of framing this state of affairs is in terms of a dissemination or implementation "problem". That is, this refusal among many clinicians to deploy ESTs is seen as stemming from ignorance or truculent resistance (Lilienfeld et al., 2013), rather than as a principled rejection of either the evidentiary hierarchy or of the general authority of scientific research.

## Conclusion

The full arc of our story starts and ends in a fundamental disagreement about the nature of evidence. In our chapter, we worked from a case study of the APA's Presidential Task Force on Evidence-based Practice to elucidate some important ways that, in recent decades, a particular view of "evidence" has been institutionalized in psychology. We outlined political factors that prompted the formation of the Task Force, institutional and practical influences that shaped the proceedings of the Task Force, and social and relational considerations that impacted deliberations within the Task Force. We also explored the ways that Task Force recommendations have been interpreted and implemented to favor a methodologically hierarchical conception of evidence; a conception not inherent in the Task Force recommendations.

Considering this story as a whole, it becomes clear that the concept of evidence in psychotherapy research and practice has been shaped by social and political influences and cannot be understood apart from these. We cannot imagine evidence as a transparent category, understood in the same ways by all, because very different notions of evidence, with very different implications and practical consequences, have been in conflict within EBP debates. There has never been a single consensual definition of evidence in psychotherapy research and practice but only a history of negotiation, persuasion, bureaucratic institutionalization, and so on, whose contours constitute what we mean by "evidence".

The fact that these contours have been more strongly shaped by a hierarchical view of evidence does not imply any consensus about the nature of evidence. Rather, the hierarchical view has exerted more influence because it is the one that has been most successfully institutionalized. This view has been built into accreditation and funding requirements, third-party reimbursement policies, and peer review practices. It has served the most purposes—like defending the disciplinary boundaries of psychology and the professional authority of clinicians.

"Evidence", in other words, has been fashioned by the requirements of those who have deployed the concept. There is some irony in this state of affairs; we don't generally see evidence as the outcome of a complex sociopolitical process, so much as a tool to arbitrate or settle debates. But, as our story here shows, to use a tool within a political contest is also to shape that tool to the needs of the contestants. There are many ways that we could think of evidence, but not all will stabilize the scientific authority of psychology; not all will carry the same weight in funding and reimbursement negotiations; not all will adapt themselves to the institutional contours of the neo-liberal manageriate. It is only those conceptualizations that will satisfy the most pressing demands within the discipline that are likely to solidify into a taken-for-granted definition of evidence.

Such political and institutional constraints will not, of course, always align with the desires, values, or intentions of the individuals caught up in them. What has been striking throughout our examination of this particular time in the history of evidence in psychology is that despite the conciliatory intentions (and recommendations) demonstrated by the Task Force, there is great diversity, including heated disagreement, in the way EBP has been implemented. Despite a manifest desire to produce a conceptualization of evidence agreeable to all, both proponents and opponents of a hierarchical view of evidence have persisted in a much more one-sided debate; and this has ultimately reduced the possibility of discourse and consensus building within the discipline.

## References

American Psychological Association. (2004). *Presidential task force on evidence based practice task force charge*. Unpublished document.

American Psychological Association. (2005a). *Council item: Policy recommendation and position paper of the 2005 presidential task force on evidence-based practice*. Unpublished document.

American Psychological Association. (2005b). *Spring consolidated meeting agenda item*. Unpublished document.

American Psychological Association. (2006a). Policy statement on evidence-based practice in psychology. *American Psychologist, 61,* 271–285.

American Psychological Association. (2006b). *Guidelines and principles for accreditation of programs in professional psychology (G&P)*. Retrieved from http://www.apa.org/ed/accreditation/about/policies/guiding-principles.pdf.

Bearman, S. K., Wadkins, M., Bailin, A., & Doctoroff, G. (2015). Pre-practicum training in professional psychology to close the research-practice gap: Changing attitudes toward evidence-based practice. *Training and Education in Professional Psychology, 9*(1), 13–20. https://doi.org/10.1037/tep0000052.

Beck, J. G., Castonguay, L. G., Chronis-Tuscano, A., Klonsky, E. D., McGinn, L. K., & Youngstrom, E. A. (2014). Principles for training in evidence-based psychology: Recommendations for the graduate curricula in clinical psychology. *Clinical Psychology: Science and Practice, 21*(4), 410–424. https://doi.org/10.1111/cpsp.12079.

Canadian Psychological Association. (2011). *Accreditation standards and procedures for doctoral programmes and internships in professional psychology* (Fifth revision). Retrieved from https://www.cpa.ca/docs/File/Accreditation/Accreditation_2011.pdf.

Canadian Psychological Association. (2012). *Evidence-based practice of psychological treatments: A Canadian perspective*. Retrieved from https://cpa.ca/docs/File/Practice/Report_of_the_EBP_Task_Force_FINAL_Board_Approved_2012.pdf.

Chambless, D. L., Sanderson, W. C., Shoham, V., Bennett Johnson, S., Pope, K. S., Crits-Christoph, P., ... McCurry, S. (1996). An update on empirically validated therapies. *The Clinical Psychologist, 49,* 5–18.

Dobson, K. S. (2016). Clinical psychology in Canada: Challenges and opportunities. *Canadian Psychology/Psychologie Canadienne, 57*(3), 211–219. https://doi.org/10.1037/cap0000061.

Eddy, D. M. (2005). Evidence-based medicine: A unified approach. *Health Affairs (Project Hope), 24*(1), 9–17.

Forman, E. M., Gaudiano, B. A., & Herbert, J. D. (2016). Pragmatic recommendations to address challenges in disseminating evidenced-based treatment guidelines. *Canadian Psychology/Psychologie Canadienne, 57*(3), 160–171. https://doi.org/10.1037/cap0000054.

Foster, E. (2015). Rivals or roomates? The relationship between evidence-based practice and practice-based evidence in studies of Anorexia Nervosa. *Counselling Psychology Review, 30*(4), 34–42.

Guyatt, G. H., Oxman, A. D., Vist, G. E., Kunz, R., Falck-Ytter, Y., Alonso-Coello, P., … The GRADE Working Group. (2008). GRADE: An emerging consensus on rating quality of evidence and strength of recommendations. *British Medical Journal, 336*, 924–926. http://dx.doi.org/10.1136/bmj.39489.470347.

La Roche, M. J., & Christopher, M. S. (2009). Changing paradigms from empirically supported treatment to evidence-based practice: A cultural perspective. *Professional Psychology: Research and Practice, 40*(4), 396–402.

Levant, R. F. (2005). Evidence-based practice in psychology. *Monitor on Psychology, 26*(2), 5.

Levant, R. F., & Hasan, N. T. (2008). Evidence-based practice in psychology. *Professional Psychology: Research and Practice, 39*(6), 658–662.

Levant, R. F., & Sperry, H. A. (2016). Components of evidence-based practice in psychology. In N. Zane, G. Bernal, F. L. Leong, N. Zane, G. Bernal, & F. L. Leong (Eds.), *Evidence-based psychological practice with ethnic minorities: Culturally informed research and clinical strategies* (pp. 15–29). Washington, DC, USA: American Psychological Association. https://doi.org/10.1037/14940-002.

Lilienfeld, S. O., Ritschel, L. A., Lynn, S. J., Cautin, R. L., & Latzman, R. D. (2013). Why many clinical psychologists are resistant to evidence-based practice: Root causes and constructive remedies. *Clinical Psychology Review, 33*(7), 883–900. https://doi.org/10.1016/j.cpr.2012.09.008.

Meehl, P. E. (1954). *Clinical versus statistical prediction: A theoretical analysis and a review of the evidence*. Minneapolis, USA: University of Minnesota Press.

Messer, S. (2004). Evidence-based practice: Beyond empirically supported treatments. *Professional Psychology: Research and Practice, 35*(6), 580–588.

McGrew, J. H., Ruble, L. A., & Smith, I. M. (2016). Autism spectrum disorder and evidence-based practice in psychology. *Clinical Psychology: Science and Practice, 23*(3), 239–255. https://doi.org/10.1111/cpsp.12160.

Rousseau, D. M., & Gunia, B. C. (2016). Evidence-based practice: The psychology of EBP implementation. *Annual Review of Psychology*, 67667–67692. https://doi.org/10.1146/annurev-psych-122414-033336.

Sackett, D. L. (2000). Evidence-based medicine. *Wiley StatsRef: Statistics Reference Online*.

Stewart, R. E., Chambless, D. L., & Stirman, S. W. (2018). Decision making and the use of evidence-based practice: Is the three-legged stool balanced? *Practice Innovations, 3*(1), 56–67. https://doi.org/10.1037/pri0000063.

Thomason, T. C. (2010). The trend toward evidence-based practice and the future of psychotherapy. *American Journal of Psychotherapy, 64*(1), 29–38.
Wampold, B. E., Goodheart, C. D., & Levant, R. (2007). Clarification and elaboration on evidence-based practice in psychology. *American Psychologist, 62,* 616–618.

# 11

# Philosophical Reflexivity in Psychological Science: Do We Have It? Does It Matter?

Kathleen Slaney, Donna Tafreshi and Charlie A. Wu

Chapter for *Psychological Studies of Science and Technology*, Edited by O'Doherty, Osbeck, Schraube, and Yen.

Given its interdisciplinary approach and emphasis on the methods by which knowledge is constructed, reflexivity is a prominent theme in much STS scholarship. Yet, for psychological science, which generally aims to identify enduring relations among psychologically relevant variables for the purpose of determining (mostly) nonhuman-contingent causes of human thought and behavior, the notion of reflexivity has not broadly penetrated its discourses or practices. Despite some recent uptake of reflexive practices within research domains where qualitative

K. Slaney (✉) · C. A. Wu
Department of Psychology, Simon Fraser University,
Burnaby, BC, Canada
e-mail: klslaney@sfu.ca

D. Tafreshi
University of the Fraser Valley, Abbotsford, Canada
e-mail: Donna.Tafreshi@ufv.ca



K. C. O'Doherty et al. (eds.), *Psychological Studies of Science and Technology*, Palgrave Studies in the Theory and History of Psychology,
https://doi.org/10.1007/978-3-030-25308-0_11

research methods are used, to the extent that reflexivity has been addressed within conventional approaches to research, it is often framed in terms of experimenter or participant bias. As such, it is viewed as something to be avoided so as not to compromise "the data." Perhaps not surprisingly, then, psychological researchers have tended not to be explicitly reflexive with regard to their philosophical commitments and how such commitments might constrain, in perhaps unintended and undesirable ways, how research is framed and approached.

However, reflexivity is a complex concept, which has been defined and used in various ways. Moreover, despite its esteemed status within qualitative research traditions, reflexivity has been a contentious issue within the broader domain of STS, with some raising the question of whether reflexivity has any utility beyond merely illuminating the processes of knowledge construction. This chapter aims to tackle this question through an examination of the contested concept of reflexivity and the various stances regarding its relevance and applicability with science, broadly, and social science, particularly. We then offer a working definition of *philosophical reflexivity*, as a subtype, and examine possibilities for a philosophical reflexive psychological science.

## Reflexivity: What, Where, and Why?

According to Smith (2005), the term reflexivity "has undoubtedly joined the pantheon of great words with multiple meanings," and, as such, "signals a cluster of debates, linked areas of inquiry, rather than a clearly articulated stance" (p. 2). Reflexivity is elsewhere characterized as: having "multiple conceptualizations" (Morawski, 2014, p. 1654); being a "contested and vaguely defined term" (Gould, 2015, p. 82) and "nebulous," "rhetorical" device (Medico & Santiago-Delefosse, 2014); "ubiquitous" and lacking "clear consensus" (Ashmore, 2015, pp. 84, 86); and suffering from an "asynchrony of definitions and procedures" (Walsh, 2003, p. 51) and much "loose talk" (Woolgar, 1988, p. 17). Thus, at the very least, reflexivity has been said to be a "family-resemblance" concept (Smith, 2005), one that is used in variety of ways, depending both on the specific domain of scholarship in which it is being applied and the particular research tradition at hand. That is to say, reflexivity is a concept that is, like most concepts, difficult to pin down in a general definition or account.

Yet, among the many definitions given of reflexivity, most are variants of the etymological definition, that is, *that which bends* or *turns back upon,* or *takes account of, itself* (Ashmore, 1989; Holland, 1999, as cited in Walsh, 2003; Lynch, 2000; Morawski, 2005; Shaw, 2010), wherein "re" means "back," and "flex" is derived from the Latin *flectere* meaning to "turn" or "bend" (Ashmore, 1989). In the context of science studies, this manifests in three related but distinguishable senses of the term. The first implies *self-reflection*, or the capacity of researchers to think about themselves in relation to the inquiries in which they engage, including their biases, theoretical predispositions, and so on (Schwandt, 2007). The second connotes *self-reference*. Here, the idea is that, within social research, subject and object are identical referents. As such, theory, by virtue of *being about* humans and their social conditions, but also *produced by* humans existing within those conditions, will have an inescapable, self-referential quality (e.g., Ashmore, 1989; Guba & Lincoln, 2005; Morawski, 2005, 2014). The third embodies a *constitutive* sense, in which reflexivity refers to the "back and forth"—that is, dynamic and mutually constitutive—relation between reality and accounts of reality (Ashmore, 1989; Morawski, 2005, 2014; Woolgar, 1988). This sense of reflexivity is often applied within ethnomethodology wherein a constructivist epistemological stance is presumed and, hence, objects of study and representations of them are not viewed as being distinct (Ashmore, 1989; Woolgar, 1988).

Given the variety of reference classes to which reflexivity is ascribed, it comes as no surprise that it should be such a polysemous concept. In some accounts, reflexivity is implied to be an *event* of one sort or another in being described variously as a process, procedure, practice, performance, act or activity. In other accounts, it is framed as an "instrument" of research, such as a form of analysis, method, or tool. Yet other treatments imply that reflexivity is a property of or relation between or among knowers and the objects of knowledge. In others, still, reflexivity is described as an ability (skill, capacity) or perspective (attitude, standpoint, argument, heuristic, critical lens) of researchers (and, in some accounts, also of research participants). More idiosyncratically, reflexivity has been described as an "ontological given" and "disciplined stance in research" (Walsh, 2003, p. 54); an "inescapable epistemological condition" and "form of human reflection" (Morawski, 2014, p. 1654); a "project of the 'self'" (Gould, 2015, p. 84); "hermeneutic reflection" (Finlay, 2003; Shaw, 2010, p. 233); an "integral data source" (Goldstein, 2017, p. 149); and "inquiry in itself" (Gemignani, 2017, p. 196).

Another source of variability in how reflexivity is conceptualized derives from the different ways in which the concept has been taken up over time, within different domains of inquiry, and for different purposes. Although it has been argued that the general notion of reflexivity plays out in many philosophical accounts of the mind and consciousness dating back at least as far as Descartes (Smith, 2005), the first consistent appearances of the reflexivity concept as it pertains to social theory came out of (primarily sociological) critiques in the 1960s and 1970s of positivism as a philosophical framework for the human and social sciences (May, 1998; Smith, 2005). An important part of the argument against positivism was that knowledge of the human domain requires explanations in terms of reasons or intentions, and not only (or even) in terms of empirical regularities, most importantly because the objects of knowledge in the human sciences, unlike those in the natural sciences, may change as a result of the knowledge scientists gain about them (Smith, 2005; Taylor, 1985). As such, accounts of reflexivity tend to be motivated by postmodern, constructivist, and hermeneutic philosophical frameworks. In the late 1980s, a small number of works dedicated explicitly to reflexivity as a central idea in the sociology of scientific knowledge (SSK) were published (e.g., Ashmore, 1989; Mulkay, 1985; Woolgar, 1988). Since then, discussion and debate about reflexivity have continued to feature heavily in STS scholarship. Curiously, outside of SSK and STS scholarship, reflexivity has been enthusiastically embraced over the past couple of decades across a broad array of social research domains in which qualitative research methods strongly feature. In fact, reflexivity is now widely viewed as an essential, and often an aspirational, component of qualitative research.

A number of explicit taxonomies of reflexivity have been proposed. From within SSK and STS scholarship, Woolgar (1988) identified the now oft-cited continuum from *radical constitutive reflexivity* to *benign introspection*. Whereas the latter assumes a distinction between accounts (representations) of the object of inquiry and the object itself, with the former, accounts (representations) and objects are thought to be intimately connected—knowledge of one entails knowledge of the other (Woolgar, 1988). Ashmore (1989) identified three senses of reflexivity commonly appearing in the various discourses within which reflexivity gets play, namely, reflexivity as *self-reference*, as *self-awareness*, and as the *constitutive circularity of accounts*. Building upon and extending Woolgar's (1988) and Ashmore's (1989) respective frameworks, Lynch (2000) proposed an inventory of reflexivities that includes *mechanical*, *substantive*, *methodological*, *metatheoretical*, *interpretative*, and *ethnomethodological*, each of which, with the exception of the last, has at least two variants. Varying accounts of reflexivity have been advanced by other well-known sociologists (e.g., Harold Garfinkel, Pierre Bourdieu, Anthony Giddens) and science studies scholars (e.g., Bruno Latour, David Bloor); together, these imply a large taxonomy of different reflexivities (see Ashmore, 2015; Wacquant, 1992 for summaries).

Several different classifications of reflexivity have also emerged from discourse on qualitative research. Wilkinson (1988) identified the three most commonly described kinds of reflexivity as *personal* (concerning the influence of the researcher's identity on inquiry), *functional* (concerning the nature of the research itself), and *disciplinary* (concerning examinations of disciplinary-specific influences on inquiry). Holland (1999) identified *personal reflexivity*, *interpersonal reflexivity*, *methodological reflexivity*, and *contextual reflexivity* (as described in Walsh, 2003). Finlay (2017) described various typologies of reflexivity that have been developed, including her own which consists of ten distinct reflexivities, namely, *introspection*, *intersubjective reflection*, *mutual collaboration*, *social critique*, *ironic deconstruction*, *strategic*, *contextual–discursive*, *embodied*, *relational*, and *ethical*. More general distinctions have also been made between *personal* and *epistemological* reflexivity (i.e., researchers' thoughts, feelings, values, experiences versus concerns directly related to research questions and methods; Hofmann & Barker, 2017); *uncritical* (e.g., bracketing) and *critical* reflexivity (Gemignani, 2017); *planned reflexive procedures* and *unplanned reflections* (Goldstein, 2017);

and reflexive practice as distinguished from reflexivity itself (e.g., Gould, 2015). As noted by Morawski (2014), the various ways of conceptualizing reflexivity encompassed by these different definitions and taxonomies hold varying implications for the degree to which reflexivity is viewed as an unavoidable epistemological condition; intentional or unintentional; an investigative problem or a tool for critically interrogating knowledge claims; as well as for how the concept is enlisted in specific realms of human affairs. We return to this point below in relation to the potential utility of philosophical reflexivity in psychological science.

Before turning to the topic of whether and how reflexivity could play a more prominent and useful role in psychological science, there are several points that first bear mentioning. These concern, respectively, the relation between the concepts of *reflexivity* and *reflection*; whether reflexivity is a problem or solution; and the differences with respect to how reflexivity features in the natural versus the social (human) sciences.

## Reflexivity Versus Reflection

The distinction between reflexivity and reflection has been the source of some consternation. This has especially been true among sociologists of science in regard to how reflexivity features in metascience study in contrast with science and other areas of knowledge production. Although the terms "reflexivity" and "reflection" (and "reflexive" and "reflective") are often used interchangeably, there appears to be general acceptance of the idea that they are not synonymous in meaning or application. At the same time, views are mixed as to the nature of their relations to one another.

For instance, Woolgar (1988) treats reflexivity and reflection ("benign introspection") as poles of a continuum—not the same but related in virtue of demarcating two extremes with respect to a range of varieties of reflexivity. Giorgi (1983, pp. 142–143) defines reflection as "bend[ing] back upon" or "tak[ing] up again" what has been pre-reflectively experienced or acted upon, which is similar to the etymological definition of reflexivity described above but not to Woolgar's definition of reflection. To reconcile these seemingly contradictory positions regarding the overlap or distinctiveness of reflexivity and reflection, it might be useful to consider Ashmore's (1989) splitting of the etymological origins of the term "reflexivity" into two primary forms: *reflect* and *reflex*. Among the discourses of *reflect* are the discourses of seeing (i.e., image of/in mirror), of thought and intellection (thinking deeply about, deliberating), and of morality (censure or reproach). The discourses of *reflex* are those of automatism (reflex action) and self-reference (reflexive grammatical form, reflexive logical relation), as well as several idiosyncratic discourses (e.g., reflex angle from geometry, Ashmore, 1989). Thus, within the discourses of seeing (mirroring), reflection connotes a similar "turning" or "bending" back as does reflexivity within the discourses of self-reference. Alternatively, within the discourses of thought and intellection, reflection connotes self-awareness and critical consideration of one's subjectivity (Goldstein, 2017) and, in the context of metascience, particularly that of "sustained self-exploration" of one's own role in and potential impact on the research (Hofmann & Barker, 2017, p. 140; Medico & Santiago-Delefosse, 2014). Such "personal" reflections (whether at the level of the individual researcher or the discipline or subdiscipline) are seen by some to be a necessary part of reflexivity, which is a broader and more critical, self-aware evaluation of the mutual influence of researcher and research (Hofmann & Barker, 2017). Within this purview, reflective practice is concerned with individual self-awareness and reflexive practice with ensuring, at the disciplinary level, that the role external, contextual factors play in informing individual subjectivity is made explicit (Evans & Hardy, 2010, as cited in Gould, 2015). However, as implied by Woolgar (1988), such perspectives on the relation between reflection and reflexivity oftentimes portray reflection as necessarily improving the adequacy of the connection between the representation and object of inquiry and, thus, maintain the postulate of the distinction between the two. Conversely, reflexivity, at least the constitutive self-referring kind, denies this distinction. Thus, both the extent to which reflection and reflexivity are presumed to work toward a common goal and the strategies for managing them vary depending on the contexts in which, and purposes for which, these concepts are taken up.

## Reflexivity: A Hindrance or a Help?

The above point dovetails nicely into the question of whether reflexivity (or, reflection, for that matter) is viewed as a problem or solution for social science. On the one hand, many general descriptions of reflexivity sing its praises for facilitating more self-critical, transparent, representative, credible, inclusive, ethical or, for some (e.g., Harding, 1992), more objective research. On the other hand, reflexivity has been blamed for complicating the scientific tasks of explanation, prediction, and control for social researchers (Flanagan, 1981). Yet, reflexivity has also been portrayed as an inescapable (Morawski, 2014) but essential (Walsh, 2003) feature of human science research. Morawski (2014) captures this tension well in her description of two overarching matters in which reflexivity finds itself: *paradox* and *irony*. The former arises from the fact that any critical consideration of reflexivity within science is itself a reflexive act and, thus, subject to the same benefits and dangers, most notably with regard to the latter, an "unending regress of reflection" (p. 1655). Moreover, within social research, observer and observed are the same ontological type and thus must somehow be differentiated to avoid self-absorption or narcissism (Banister, 2011; de Saint-Laurent & Glaveanu, 2016; Maton, 2003); solipsism or relativism (Finlay, 2017; Smith, 2005); or rhetorical problems (Sismondo, 2010). Particularly contentious is the notion that the strong "X of X" structure of SSK, with its attendant relativist epistemology, falls seriously prey to the self-refutation argument (i.e., if scientific knowledge does not constitute truth, generally, then knowledge about reflexivity produced within SSK, and STS more broadly, also does not constitute truth; Ashmore, 2015). Yet, because it is an unavoidable feature of social inquiry, reflexivity invokes an irony whenever social scientists disregard or dismiss it as an essential attribute of the human condition and thus a means to knowledge about the human condition (Morawski, 2014).

There has been a range of responses to the questions of whether reflexivity is a problem, and, if so, for whom, and how it should be managed. However, speaking very generally, one might say reflexivity *as critical reflection* about the practices of and knowledge claims resulting from scientific inquiry (both the natural and human varieties) is viewed as a desideratum of "good" (valid, ethical) science. Yet, in the human sciences, variants of constitutive reflexivity must be acknowledged and framed accordingly in the accounts given by social researchers such that we gain greater understanding of the particulars with respect to the scope and validity of social theory (Taylor, 1985). Moreover, how "radical" reflexivity needs to be depends to a large extent on whether it is called upon to examine the construction of knowledge (scientific claims) versus the construction *of* the construction (SSK claims) (Lynch, 2000).

## Reflexivity in the Natural Versus the Social Sciences

Flanagan (1981) claimed that reflexivity is the "favorite candidate for the property which makes the human sciences unique" because it involves "the unique capacity of humans to engage in self-conscious inquiry into their own condition" and, in so doing, informs us about the nature of the "entities" (i.e., humans) being studied—that is, that they are capable of reflecting (p. 375). Put simply, whereas quarks, genes, and chemical compounds cannot reflect upon their own natures and conditions of existence, humans can, and this capacity is a fundamental part of what it means to be human. The social and natural sciences have also been distinguished in terms of reflexivity by virtue of the different implications theorizing has for the objects under study. Whereas social theories can undermine, strengthen or shape the human practices they bear on, in the natural sciences, theory pertains to independent objects[1] (Taylor, 1985). Another means of distinction that hinges on reflexivity is that, whereas both the natural and social sciences may (potentially) involve introspective reflexivity (i.e., reflection upon

[1]Notwithstanding implications for the notion of observer independence (Popoveniuc, 2014; Smith, 2005), which further complicates this issue. However, discussion of this thorny topic is well outside of the scope of the current work.

the presuppositions and activities of the science) and are interpretive in the sense that knowledge is constrained by the contextualizing frames within which scientists conduct their inquires, the latter involves a "double-hermeneutic." That is, social researchers both are shaped by the societies of which they are a part and contribute to shaping those societies through their theorizing (Bishop, 2007; May, 1998; Taylor, 1985). A third distinction that has been made by Smith (2005) concerns the potentially different purposes served by reflexive analysis in the natural and social sciences, respectively. Whereas reflexive analysis in natural science inquiries primarily serves to reveal underlying assumptions that attend and potentially influence knowledge production about objects and relations whose existence does not depend on humans' awareness or understanding of them, reflexivity in the social sciences itself has the potential to foster a self-conscious awareness, and hence, knowledge of the social realities of the human objects under study.

## Reflexivity and Psychological Science

Psychology has often been described as standing apart from most other social research disciplines in virtue of its persistent resistance to or abeyance of reflexivity, or worse, its outright denial of it (Morawski, 2005, 2014). According to Morawski (2005), with but a few notable exceptions, disregard for the problem of reflexivity has been a feature of psychological science from its earliest history. In fact, within the emerging science of psychology in the late nineteenth century, scientists' reflection upon their place in research was eschewed on the grounds that it threatened to contaminate experimental procedures and, consequently, the status of psychology as a science. As a result, unlike in some domains of social science (e.g., sociology, feminist and gender studies), there has been no disciplinary level development within psychological science of critical reflexive practices. That is to say, generally speaking, psychological researchers have not engaged in critical examinations of their positions as observers and producers of knowledge (Morawski, 2005). Rather, if acknowledged at all, reflexivity is something to be controlled or eliminated through the putatively objective, rigorous, and precise methods that are privileged within the discipline (Morawski, 2014;

Tafreshi, Slaney, & Neufeld, 2016). Moreover, challenges from feminist and other critical and theoretical psychologists to recognize the epistemological problems of reflexivity have largely been contained to the margins and have thus failed to penetrate mainstream theory or practice (Morawski, 2014; Smith, 2005).

Although sustained examinations of reflexivity as an epistemological challenge have been rare within psychology, within the past couple of decades there has been growing interest among qualitative researchers in incorporating reflexive analysis into research practice (Henwood, 2008). With the increasing profile and popularity of qualitative research within psychology (Gergen, Josselson, & Freeman, 2015), avenues have begun to open for embracing reflexivity as an important component of human science. However, although movement toward psychological researchers thinking critically about their role within and impact on research might be considered by some to be a sign of progress, concerns have been raised about the potential dangers of adopting reflexivity in an uncritical—or, shall we say, *unreflective*—way (Gemignani, 2017). At best, such "benign" introspections limit reflexivity to a perfunctory component of research procedure; at worst, reflexivity is reduced to a kind of confessional for individual researchers to name their biases and, as such, may contribute to reducing structural problems to personal ones, individualize the social, political, and moral components of science, and limit consideration of reflexivity to methodological intervention (Burman, 2006, as cited in Teo, Gao, & Shevari, 2014). In the following section, we outline one potential way in which reflexivity might be adopted within psychological science. In particular, we focus on possibilities for a philosophically reflexive science of psychology and consider whether, and the extent to which, the etymological senses elaborated in the first part of the current chapter will be relevant under this proposal.

## Philosophical Reflexivity: A Special Breed?

There is a trivial sense in which all meanings of reflexivity are philosophical, in that all tend to be articulated at least to some extent with philosophical language or framed within broader philosophical, especially epistemological, topics. Moreover, as noted, some have aligned reflexivity with philosophical traditions such as hermeneutics and phenomenology (e.g., Finlay, 2003; Shaw, 2010; Walsh, 2003) or with postmodern and postconstructivist epistemologies (e.g., Gemignani, 2017; Tuval-Mashiach, 2017). However, what we mean by philosophical reflexivity is somewhat more general in that, similar to Teo et al. (2014), we construe reflexivity in terms of broader ontological, epistemological, and ethical dimensions. That is, first, philosophical reflexivity involves critical thinking about the nature of the objects under study, that is, the kinds of "things" they are and the kinds of relations into which they enter. Of course, for psychological researchers and scholars, this is no simple issue and there has been longstanding debate as to whether psychology should be construed as a science or a humanity, with strong advocates for each position and several grades in between (see Koch, 1981; Korn, 1985; Teo, 2017). Certainly, we are not going to attempt to settle this issue here. However, we believe that, at the very least, it could be conceded that psychology, as a domain of inquiry and practice, is ontically plural. That is, it is concerned with more than one ontological category and thus more than one sense of "real" (Slaney & Tafreshi, 2019). Such pluralism implies a range of subject matters falling under the broad umbrella of psychology, concerning everything from predicting event-related potential waveforms in neural activity to probing the impact of neoliberal ideology on the formation of persons. Accordingly, a philosophically reflexive psychology requires, minimally, critical reflection upon the nature of the objects of study in a given research project or body of work.

A second overarching feature of a philosophically reflexive psychology is an openness to epistemic plurality in response to recognition that the specific natures of the objects under study imply varying epistemological constraints and, hence, knowledge of the "psychological" cannot be neatly packaged into categories such as "objective," "subjective," "constructed," "interpretative," "relativistic," and so on. Both ontic and epistemic plurality give rise to the need for methodological pluralism, which ideally requires reflection upon and a general openness to considering from a broad array of methods and choosing those that are best suited for the purposes of a given inquiry within the context in which it is conducted (Slaney & Tafreshi, 2019).

A third feature of philosophical reflexivity concerns the ethical and moral dimensions of scientific research. Of course, it is generally now accepted among philosophers of science that, as a human activity that involves making judgments at every stage of research, all science involves values and, thus, produces knowledge that is value-laden, at least to some extent. Moreover, whereas the "value-free ideal" at the center of debates about values in science in the mid-twentieth century relied on a firm distinction between so-called "epistemic" (related to knowledge, and thus "acceptable") and "non-epistemic" (related to social, ethical, and other factors, and thus "forbidden"), more recent examinations of values in science eschew so a crude a distinction and call for a more "nuanced topography" of values in science (Douglas, 2009).[2]

There is an enormous literature on the axiological dimension of research, a summary of which will not be attempted here. Suffice it to say that, like all sciences, psychological research is inherently social, and, so too, inherently moral (Teo et al., 2014). The ethical–moral dimension of research in psychology and other social sciences is complicated further by the fact that the objects under study are creatures who themselves hold values, which will not necessarily align with those that motivate or constrain the judgments of and decisions made by scientists. Moreover, a good deal of psychological research actually involves an examination of value-laden social practices (not just those of the scientist) and, therefore, involves researchers giving self-descriptions of those practices, at least some understanding of (and reflection upon) is implicit in those self-descriptions (Taylor, 1985). A philosophically reflexive psychological science, thus, would require researchers and communities of researchers to recognize and identify the types of values and other moral contents of the subject matters that are the focus of their inquires and inform their research practices in light of the moral and ethical implications of all components of the research in which they engage.

[2]For example, Douglas (2009) identifies different types of scientific values (ethical, social, and cognitive, and epistemic) and different roles (direct and indirect) values play within science. Further, Douglas argues that so-called epistemic "values" are more like virtues than values, in the sense that they function as criteria that should, ideally, be met by scientific theories.

## Philosophical Reflexivity in Light of the Etymological Definition

As we have described above, reflexivity is a textured concept that has a range of applications in a variety of discourses. However, for the purpose of drawing out implications for what we envision as a potentially useful philosophical reflexivity for scientific psychology, we believe it is useful to narrow in on the three general variants of the etymological definition described above. It is also important to keep in mind the distinction between critical and uncritical reflexivity, as well as between reflexivity as a desired and intentional practice within science versus as an inevitable feature of (scientific) knowledge production. We briefly address these themes and make some allusions to potential methodological implications for a philosophically reflexive psychology.

Recall that most senses of the term "reflexivity" (and also "reflection") connote a turning or bending back upon, or taking account of, oneself (Ashmore, 1989). Recall also that this root meaning manifests in three related but distinguishable senses: self-reflection, self-reference, and self-constitution. The question here is whether, and if so, the extent to which, each of these senses comes into play when reflexivity is constrained to the sort of philosophical reflexivity the current chapter is concerned with.

With respect to the *self-reflective* sense, as noted, philosophical reflexivity involves researchers incorporating into their praxis critical reflection upon what kinds of objects and events are under study (ontological considerations), what boundaries and constraints exist regarding our knowledge of them (epistemological considerations), the moral and ethical landscape of the research (axiological considerations), and, in light of all of these, what are the best, or at least better, tools for conducting sustained inquiries about them (methodological considerations).

Of course, in no way do we intend to suggest that ontological, epistemological, and axiological domains are neatly separable or hierarchically structured in a clear way. They certainly are not, and the boundaries among them are much fuzzier for some types of inquiry than others. Nor do we wish to suggest that critical self-reflection, on the part of individual psychological researchers or the discipline as a whole, carries the burden of dogmatic ontological (e.g., subjectivist), epistemological (e.g., constructivist) or methodological (e.g., phenomenological, activist-oriented) commitments. Our goal is not to advance any particular perspective on where and how the ontological, epistemological, and axiological landscapes of science make contact. Rather, our aim is to argue that they *do* make contact and thus to advocate generally for a philosophical reflexivity in psychology that involves sustained critical examinations of the ontic, epistemic, and moral–ethical "spaces" that constrain and enable psychological research in various ways.

Given the ontic and epistemic plurality, and range of ethical–moral considerations, encountered within the broader discipline of psychology, a careful consideration of what kinds of "things" and what kinds of relations such "things" are capable of entering into is an essential first step for researchers working within a philosophically reflexive psychology. Then, meaningful critical examination of the epistemic conditions of knowledge production for the types of inquiry at hand is possible. Both ontological and epistemological considerations carry ethical implications for real and potential consequences of inquiry (Teo et al., 2014). In this respect, a philosophically reflexive psychology involves scrutiny of potential ethical repercussions—for research participants, as well as stakeholders and consumers of research—of everything from the research question itself, influences of the researcher on the research and of both on research participants, methods used, to knowledge transmission and application. This entails a *self-accounting* sense of self-reflective philosophical reflexivity. It affords the opportunity for psychological researchers to examine where psychology—or a particular subfield, research program, or research question—is situated among the sciences, and among scholarly disciplines more broadly. It also involves critical examination of the moral and ethical dimensions of the human condition, which should cycle back through ontological and epistemological considerations to both self-referential and self-constitutive senses of reflexivity.

A philosophically *self-referential* reflexivity is less *reflective*—in a self-examination or self-accounting sense—and more *reflexive*, in the sense that, like all other domains of social research, psychology is both about and conducted by humans.[3] Morawski (2005) frames this as the "inescapable" self-referential quality of theory in the social sciences. However, this need not imply a self-defeating relativism at the level of scientific (as opposed to metascientific) discourse (Ashmore, 1989). Moreover, although it certainly bears on practice, this aspect of a philosophically reflexive psychology simply acknowledges epistemological constraints as a feature of much psychological research. However, as Gould (2015) notes, the general epistemological condition of self-referential reflexivity possessed by all humans is most often tacit but "can be transmitted and refined into higher levels of competence" within research practice (Gould, 2015, p. 85).

The sense in which psychology is *constitutively* reflexive is the most complex and has the greatest implications for current research conventions. The disanalogy between the natural and social sciences highlighted by Taylor (1985), and alluded to above, is important here. Whereas natural science theories bear primarily on ontically independent objects, social theories (of which, it could be reasonably argued, the vast majority of psychological theories consist) do not: that is, they are self-referential and bear on human practices, the latter of which are partly constituted by self-understandings. As such, unlike natural science theories, social theories "can undermine, strengthen or shape the practice they bear on" and, in so doing, transform self-understandings, thus potentially altering the practices, and so on[4] (Taylor, 1985, p. 101). Put simply, for many areas of psychological research, both our participation in the social world at large and how we as a social group (i.e., of scientists) approach and manage our inquiries produces in part how we come to understand ourselves as certain kinds of psychological beings (or beings with a certain sort of psychology).[5] This aspect of philosophical reflexivity, it could be argued, has more profound implications for current practice than either the self-reflective or self-referential

[3]With the exception of research that involves animal subjects and/or pertains to animal psychology.

[4]Of course, this also pertains to social theories *about* science and bears on some of the more contentious debates about reflexivity among sociologists of science and other meta-science scholars.

[5]The works of Kurt Danziger and Ian Hacking are particularly notable with respect to this point.

aspects. This is primarily because the standard representational frameworks, which presume a clear subject-object divide, will be wholly insufficient for capturing the "complex dialectic between knower and known" (Finlay, 2003, p. 235). Instead, a versatile methodology and set of tools will be required, including a wide variety of qualitative methods, and a well-honed practice of critical reflection in order to choose the best suited among available methods for the inquiry at hand. However, as with self-referential reflexivity, we deny the postmodern stance that constitutive reflexivity necessarily implies that no account or method can be valued over any other (Shaw, 2010).

## Conclusions

In closing, it might be fair to ask whether fostering philosophical reflexivity in psychology will matter all that much at the levels of either theory or practice. Certainly some STS scholars have raised the specter of doubt on that front (e.g., Bausch [2002] argues that a scientific work can be useful even in the absence of philosophical reflexivity; Lynch [2000] argues the "epistemological hubris" that often accompanies self-consciously reflexive claims, though sometimes insightful, is often unwarranted; see also Ashmore, 2015). Clearly, by virtue of asking the question in the present chapter, we embrace reflexivity, at least the self-reflective variety. Yet, we acknowledge that the prescriptions we offer here are general and nonspecific and, as such, run the risk of being too vague to be of any concrete utility. It is not the objective of the present work, however, to provide a "how to" guide in being reflexive for psychological researchers. Such an attempt would fly in the face of the kind of philosophical reflexivity we advocate for, from the perspective of which the particulars regarding the nature of the phenomena under study, the relevant epistemological constraints, and the ethical–moral issues at play need to be worked out on a case-by-case basis. Our objective here has been, simply, to illuminate the potential benefits (but, also, some of the dangers) of reflexivity within psychological science in contrast to the current state of affairs in which philosophical blindness and theoretical and methodological agnosticism prevail. While obviously not the answer to all of psychology's woes, we believe that endorsing philosophical reflexivity is a step in the right direction toward a more critical, methodologically plural, transdisciplinary, and socially just framework for handling the broad and diverse set of inquiries that collectively make up the discipline of psychology.

## References

Ashmore, M. (1989). *The reflexivity thesis: Wrighting sociology of scientific knowledge*. Chicago: University of Chicago Press.

Ashmore, M. (2015). Reflexivity in science and technology studies. In *International encyclopedia of the social and behavioral sciences* (2nd ed., Vol. 20, pp. 77–93). https://doi.org/10.1016/b978-0-08-097086-8.85018-7.

Banister, P. (2011). *Qualitative methods in psychology: A research guide*. Maidenhead: Open University Press.

Bausch, F. A., Jr. (2002). *Examining one's own: Reflexivity and critique in STS*. Unpublished masters thesis.

Bishop, R. C. (2007). *The philosophy of the social sciences*. London: Continuum International Publishing Group.

de Saint-Laurent, C., & Glaveanu, V. (2016). Reflexivity. In V. P. Glaveanu, L. T. Pedersen, & C. Wegener (Eds.), *Creativity: A new vocabulary* (pp. 121–128). London: Palgrave Macmillan.

Douglas, H. E. (2009). *Science, policy, and the value-free ideal*. Pittsburgh, PA: University of Pittsburgh Press.

Finlay, L. (2003). Through the looking glass: Intersubjectivity and hermeneutic reflection. In L. Finlay & B. Gough (Eds.), *Reflexivity: A practical guide for researchers in health and social sciences* (pp. 103–119). Oxford, UK: Blackwell.

Finlay, L. (2017). Introduction: Championing "reflexivities". *Qualitative Psychology, 4,* 120–125.

Flanagan, O. J., Jr. (1981). Psychology, progress, and the problem of reflexivity: A study in the epistemological foundations of psychology. *Journal of the History of the Behavioral Sciences, 17,* 275–386.

Gemignani, M. (2017). Toward a critical reflexivity in qualitative inquiry: Relational and posthumanist reflections on realism, researcher's centrality, and representationalism in reflexivity. *Qualitative Psychology, 4,* 185–198.

Gergen, K. J., Josselson, R., & Freeman, M. (2015). The promises of qualitative inquiry. *American Psychologist, 70,* 1–9.

Giorgi, A. (1983). Concerning the possibility of phenomenological psychological research. *Journal of Phenomenological Psychology, 14,* 129–169.
Goldstein, S. L. (2017). Reflexivity in narrative research: Accessing meaning through the participant-researcher relationship. *Qualitative Psychology, 4,* 149–164.
Gould, N. (2015). Reflexivity. In *International encyclopedia of the social and behavioral sciences* (2nd ed., Vol. 20, pp. 82–87). https://doi.org/10.1016/b978-0-08-097086-8.28075-6.
Guba, E. G., & Lincoln, Y. S. (2005). Paradigmatic controversies, contradictions, and emerging confluences. In N. K. Denzin & Y. S. Lincoln (Eds.), *Handbook of qualitative research* (3rd ed., pp. 191–215). Thousand Oaks, CA: Sage.
Harding, S. (1992). Rethinking standpoint epistemology: What is "strong objectivity?" *The Centennial Review, 36,* 437–470.
Henwood, K. (2008). Qualitative research, reflexivity and living with risk: Valuing and practicing epistemic reflexivity and centering marginality. *Qualitative Research in Psychology, 5,* 45–55.
Hofmann, M., & Barker, C. (2017). On researching a health condition that the researcher has also experienced. *Qualitative Psychology, 4,* 139–148.
Koch, S. (1981). The nature and limits of psychological knowledge: Lessons of a century qua "science". *American Psychologist, 36,* 257–269.
Korn, J. H. (1985). Psychology as a humanity. *Teaching of Psychology, 12,* 188–193.
Lynch, M. (2000). Against reflexivity as an academic virtue and source of privileged knowledge. *Theory, Culture & Society, 17*(3), 26–54.
Maton, K. (2003). Reflexivity, relationism, and research: Pierre Bourdieu and the epistemic conditions of social scientific knowledge. *Space and Culture, 6*(1), 52–65.
May, T. (1998). Reflexivity in the age of reconstructive social science. *International Journal of Social Research Methodology, 1,* 7–24.
Medico, D., & Santiago-Delefosse, M. (2014). From reflexivity to resonances: Accounting for interpretation phenomena in qualitative research. *Qualitative Research in Psychology, 11,* 350–364.
Morawski, J. (2005). Reflexivity and the psychologists. *History of the Human Sciences, 18*(4), 77–105.
Morawski, J. (2014). Reflexivity. In T. Teo (Ed.), *Encyclopedia of critical psychology: Springer reference* (pp. 1653–1660). New York: Springer.
Mulkay, M. (1985). *The word and the world: Explorations in the form of sociological analysis.* London: George Allen & Unwin.

Popoveniuc, B. (2014). Self reflexivity: The ultimate end of knowledge. *Procedia-Social and Behavioral Sciences, 163*, 204–213.

Schwandt, T. A. (2007). *The Sage dictionary of qualitative inquiry*. Thousand Oaks, CA: Sage. https://doi.org/10.4135/9781412986281.

Shaw, R. (2010). Embedding reflexivity within experimental qualitative psychology. *Qualitative Research in Psychology, 7*, 233–243. https://doi.org/10.1080/14780880802699092.

Sismondo, S. (2010). *An introduction to science and technology studies* (2nd ed.). Chichester, West Sussex, UK: Wiley-Blackwell.

Slaney, K. L., & Tafreshi, D. (2019). Quantitative, qualitative, or mixed? Should philosophy guide method choice? In B. Schiff (Ed.), *Situating qualitative methods in psychological science* (pp. 27–42). New York: Routledge.

Smith, R. (2005). Does reflexivity separate the human sciences from the natural sciences? *History of the Human Sciences, 18*(4), 1–25. https://doi.org/10.1177/0952695105058468.

Tafreshi, D., Slaney, K. L., & Neufeld, S. D. (2016). Quantification in psychology: Critical analysis of an unreflective practice. *Journal of Theoretical and Philosophical Psychology, 36*, 233–249.

Taylor, C. (1985). Social theory as practice. In *Philosophy and the human sciences: Philosophical papers* (Vol. 2). Cambridge: Cambridge University Press.

Teo, T. (2017). From psychological science to the psychological humanities: Building a general theory of subjectivity. *Review of General Psychology, 21*, 281–291.

Teo, T., Gao, Z., & Sheivari, R. (2014). Philosophical reflexivity in social justice work. In C. V. Johnson & H. Friedman (Eds.), *The Praeger handbook of social justice and psychology* (pp. 65–78). Santa Barbara, CA: Praeger.

Tuval-Mashiach, R. (2017). Raising the curtain: The importance of transparency in qualitative research. *Qualitative Psychology, 4*, 126–138.

Wacquant, J. D. (1992). Toward a social praxeology: The structure and logic of Bourdieu's sociology. In P. Bordieu & J. D. Wacquant (Eds.), *An invitation to reflexive sociology* (pp. 1–60). Chicago: University of Chicago Press.

Walsh, R. (2003). The methods of reflexivity. *The Humanistic Psychologist, 31*(4), 51–66.

Wilkinson, S. (1988). The role of reflexivity in feminist psychology. *Women's Studies International Forum, 11*, 493–502.

Woolgar, S. (1988). Reflexivity is the ethnographer of the text. In S. Woolgar (Ed.), *Knowledge and reflexivity: New frontiers in the sociology of knowledge* (pp. 14–34). London: Sage.

# 12

# A Meeting of Minds: Can Cognitive Psychology Meet the Demands of Queer Theory?

**Sapphira R. Thorne and Peter Hegarty**

## Introduction

For over fifty years, cognitive psychologists have grappled with how best to understand how people conceptualize. Since the cognitive revolution, concepts have been defined in diverse ways that tend to assume that they form the 'building blocks' of abstract rational thought (Solomon, Medin, & Lynch, 1999). Theories of categorization are diverse and have been narrated as successive waves of categorization research from *classical* (categories with discrete boundaries) to *probabilistic* (categories formed by prototypes and exemplars) to *explanation-based* (categories based on explanations), with each wave demonstrating how the former

S. R. Thorne (✉)
School of Psychology, Cardiff University, Cardiff, UK
e-mail: ThorneS5@cardiff.ac.uk

P. Hegarty
School of Psychology, University of Surrey, Guildford, UK
e-mail: p.hegarty@surrey.ac.uk



K. C. O'Doherty et al. (eds.), *Psychological Studies of Science and Technology*, Palgrave Studies in the Theory and History of Psychology,
https://doi.org/10.1007/978-3-030-25308-0_12

wave failed to account for the complexity and flexibility of how humans reason with categories (Hampton, 2010; Komatsu, 1992; Margolis, 1994; Medin, 1989). In this chapter, we take a different stance to the study of categorization and make an argument about the affordances of cognitive theories of categorization for queer approaches to science and technology.

From the vantage point of critical psychology, this project may seem strange. A traditional view in social cognition is that people categorize others by race, gender, and other matters automatically and uncontrollably, and that these categories allow people to operate as 'cognitive misers' spending little resources making sense of others (e.g., Taylor, Fiske, Etcoff, & Ruderman, 1978). From this perspective, queering psychology seems like an impossible task that asks people to do something very unnatural. In this chapter we argue that thinking queerly might be much easier than the 'cognitive miser' view allows, and we use the categorization systems developed by cognitive psychologists in order to make our argument.

We mean this essay to do something akin to Barad's notion of *diffractive reading*; 'a transdisciplinary reading approach that remains rigorously attentive to important details of specialized arguments within a given field, in an effort to foster constructive engagements across (and a reworking of) disciplinary boundaries' (p. 25). As such, we first examine what is at stake in queering psychology, next review the history of cognitive psychological studies of concepts, and lastly draw both together to discuss contemporary critical psychological work in intersex.

## Queer Theory and Psychology

Queer theory emerged in the late 1980s and early 1990s as a refusal to conceptualize lesbian, gay and queer experiences primarily in terms of their difference from heterosexuality, but instead to make sense of them as cultures in their own right (Jagose, 1996; Turner, 2000). Strongly influenced by the dark visions of psychology, psychiatry, and sexology in the work of Michel Foucault (1976/1998) queer theorists tended to assume that psychology was a means for the exercise of disciplinary power, which worked not so much by rendering people invisible, but by creating documentation about them (Foucault, 1975). Foucault urged distrust of appeals to 'natural' sexuality, insisting that all ideas of the 'natural' were grounded in conceptual frameworks produced first-and-foremost to legitimate the exercise of power (Foucault, 1976/1998). Queer theorists refused to posit what queer theory was for, preferring instead to describe what is negated:

> Queer is by definition whatever is at odds with the normal, the legitimate, the dominant. There is nothing in particular to which it necessarily refers. It is an identity without an essence... a positionality that is not restricted to lesbians and gay men but is in fact available to anyone who is or who feels marginalised because of her or his sexual practices. (Halperin, 1995, p. 62)

Queer theory was, as such, more explicit about negation than affirmation.

For this reason, queer theory seemed at odds with psychology's liberal narrative of 'affirmation' of lesbian and gay identities, and later of bisexual and transgender identities (Clark, Ellis, Peel, & Riggs, 2010; Downing & Gillett, 2011; Johnson, 2015; see Hegarty, 2017 for a history). One response to the representational limitations of affirmative particular social groups is to understand queer not as one-more identity, but as shifting the structure of the category to that of an umbrella term. It is not precisely clear what such an umbrella term covers, but it is oriented toward sheltering diverse people whose primary shared attribute may be their vulnerability to heteronormativity and related forms of epistemological violence (Teo, 2010).

There may be more freedom in a category so defined, than in one defined in terms of the possession of fixed social identities. Social categories do more than identify. They can also create strict boundaries, such as those that demarcate female and male, straight and gay, etc. This very ambiguity of queer—as both a named space under the umbrella and a comment on the umbrella's structure—suggests the ordinariness with which people make generous sense of the world in non-miserly ways.

Several mainstream developments in LGBT psychology in the early 2000s, such as research on sexual fluidity (Diamond, 2003, 2004), the political meanings of essentialist beliefs (Hegarty, 2002), and historically situated life narratives of lesbians, gay men and bisexuals (Hammack & Cohler, 2009), were influenced by queer theory. Critical psychology and lesbian and gay affirmative psychology presented dilemmas for each other about engaging mainstream psychology's liberal vision of advancing equal rights and engaging in psychology's positivist-empiricist epistemologies (see Kitzinger, 1997 for an early discussion). Social constructionism explicitly opposed positivist-empiricist assumptions that something is either A or B, and that empirical research should determine which of those two things it is (e.g., Brown, 1989; DeLamater & Hyde, 1998). If psychology is to be queer, then it must be qualitative in the minds of some (Warner, 2004). For others, the only obvious reason to engage in quantitative psychology was to establish matters of fact, to achieve some kind of political end (e.g., Kitzinger & Coyle, 2000; Rivers, 2000). However, so doing, overlooks the politics of fact-making practices and particularly the fact that human research subjects routinely do things agentially that surprise and confound researchers, some of which are captured as 'data' and some of which elude such categorization (Hegarty, 2001, 2007). As approaches such as agential realism make clear, it is possible to engage in forms of experimentation that are directly tied to philosophical and political questions, and which presume that phenomena A and B do not preexist experimental observation, but become stable, replicable consequences of observation in interaction with experimenters' and their tools.

## Concepts and Categories: A Short Review

Cognitive psychology was strongly influenced by cybernetics and Gestalt theories which emphasized the active constructive properties of minds in formulating hypotheses (Tolman, 1948; Wason, 1960). The positivism critiqued by social constructionists has roots in a 'classical' view of categorization, a view that assumes that categories have discrete boundaries defined by their necessary and sufficient properties which scientific observation can discern (Popper, 1959/2002). Cognitive psychology's epistemology is informed by Popperian norms for scientific logic. By this we mean that cognitive psychologists conduct experiments that subject existing theories of categorization to tests of falsification; theories that survive tests of falsification are accepted while those that are falsified by evidence are rejected. This epistemology presumes that things exist or not prior to their becoming objects of study, putting cognitive psychology at odds with some versions of social constructionism that emerged in sexuality research (DeLamater & Hyde, 1998). We will return to the question of what cognitive psychologists can be described as having done, in relation to these logical Popperian norms, in the conclusion. In the interim, our reading of its history aims to diffract cognitive psychology's findings and its logical commitments.

Research conducted prior to the 1970s endorsed the *classical* theory that people represent concepts with necessary features, possessed by every member, and sufficient features, possessed only by category members, which jointly render categories quite clear-cut and discrete (Machery, 2009; Medin, 1989). The social category bachelor was often used to exemplify the classical theory, on the ground a bachelor possesses the necessary, and binary, features of being unmarried and a man. While Popperian logic requires drawing out such classical implications of theories and testing them, reasoning as if categories were classical creates the conditions for the mis-recognition of people who have a mix of necessary and sufficient features (Dunham & Olson, 2016), as appears to be the case commonly with gender (Hyde, Bigler, Joel, Tate, & van Anders, 2018). The example of bachelor also shows the limits of the classical theory, as both gender and marital status are normative concepts with additional teleological meanings, that are historically specific. These meanings are evident in widespread social anxieties that bachelors might not end up married which characterized the nineteenth century moment when the word became widespread (Bertolini, 1996). The movement toward an increased recognition of diversity in terms of gender (e.g., intersex) and marital status (e.g., civil partnership) in the twenty-first century renders this example problematic.

In recognition of some of the shortcomings of classical theory, Eleanor Rosch's *prototype theory* allowed concepts to be represented by attributes that share a *family resemblance* (Rosch, 1975; Rosch & Mervis, 1975; see Mervis & Rosch, 1981). The metaphor of family resemblance—drawn from Wittgenstein (1953)—moved cognitive psychologists to imagine that the possession of similar naturally occurring attributes explained category membership. Rosch's prototype view presented category membership as continuous based on the possession of central and peripheral features that were present in a category *prototype*, which represented a category's most typical member. Thus a robin was more central to the category bird than an ostrich, as it possesses more similar features to the prototype, and often-replicated experiments showed that category centrality eases processing, learning, and retrieval (Rosch, 1973, 1975; Rosch, Simpson, & Miller, 1976). Rosch's work pertained to object categories, such as furniture, animal categories, such as birds, and was captured by debates about the naturalness of color categories, whose labels vary considerably between human languages (Berlin & Kay, 1969). We emphasize three critical directions of categorization research that followed from Rosch's work next.

First, consider how critical scholars in science and technology studies know that humans usually categorize for some kind of social purpose, and that they often reason with classical and prototype representations of categories at once (Bowker & Starr, 1999). Goal-directed categories such as 'foods to eat on a diet' rarely have a 'family resemblance' and their best examples are *ideal* types rather than averages (Barsalou, 1982, 1983). This point seems to echo Butler's (1990) theory of gender, among others, as a system that is put in place to achieve certain ends, and which creates the sense not of what genders are, but what they *might* or *should be*. The goals that people have for creating their categorizations vary between individuals, between cultures, and by expertise (see Medin et al., 2006). Moreover, Barsalou's (1982) example of 'foods to eat on a diet' touches particularly on the normalization of gender. Dieting and body image are areas where the experience of reality and ideals differ (Fallon & Rozin, 1985), and the line between caring for oneself and subjecting oneself to normalization requires Foucaultian insights to discern (Heyes, 2006).

Second, to the extent that prototype theory assumes relatively stable graded structure it struggles to explain contextual variability in meaning

(Roth & Schoben, 1983). People routinely turn the volume up or down on particular social category memberships in ordinary social situations all the time (Tajfel & Turner, 1986). For example, in some situations our behavior is interpreted by ourselves and our interaction partners as an expression of gender and in other situations it is not (Deaux & Major, 1987). Failing to consider the ordinariness with which people change categories to fit social contexts can not only underestimate, but curtail the agency that people need to exercise in ordinary life.

Third, the meaning of categories changes when such categories are combined (e.g., Hampton, 1987; Hastie, Schroeder, & Weber, 1990; Murphy, 1988; Smith & Osherson, 1984; Smith, Osherson, Rips, & Keane, 1988). In the 1990s, researchers started to consider how social categories may be combined to produce new categories (e.g., see Hastie et al., 1990; Kunda, Miller, & Claire, 1990; Kunda & Oleson, 1995; Storms, De Boeck, Van Mechelen, & Tuts, 1996). Categories of people under the queer umbrella often emerged as good examples of category combination effects. For example, stereotypes of a 'gay construction worker' that come to mind might not have the prototypical features of either a gay man or a construction worker as prototype theories would presume. Rather, novel meanings emerge from constructing a narrative in response to that surprising conjunction, such as one participant in Kunda et al. (1990, p. 556) who constructed the narrative that 'This person is most likely sublimating.' This facility for narration of category combinations was also demonstrated some years earlier by Kessler and McKenna (1978). They delivered a series of preprepared random yes/no answers to questions that might allow participants to discern the gender of a person that the experimenter held in mind. Participants were quick and successful in resolving the discrepancies they produced (reasoning that a person with a beard in a skirt might be a Hawaiian or Scottish man, for example), but always by reiterating the logic of the two-gender system in making sense of such cases. These experiments should be read in light of Jerome Bruner's (1990) arguments at this time that the basic unit of sense-making was not information but narrative.

Inflexibility in the prototype view prompted a number of 'exemplar' theories of categorization. Exemplar theory was developed as an umbrella term to refer to theories which assumed that conceptual structure was defined by multiple exemplars rather than a single prototype (e.g., Medin & Schaffer, 1978; Smith & Medin, 1981). Prototypes may describe the conceptual structure of novices, while exemplars describe the conceptualizations that experts use in a domain (e.g., Genero & Cantor, 1987).

One theory of category activation relied on both prototype and exemplar representation to approach a question that is central to queer inquiry; how do people think in ways that assume that some events or people are the taken-for-granted norm? Kahneman and Miller's (1986) norm theory assumes that the categories used to think with have a prototype structure, but also that different exemplars of the same category can be activated to form a working representation or norm that supports abstract thought about a category or event. Consider for example the account of norms offered by Lorde (1984, p. 116):

> Somewhere on the edge of consciousness there is what I call a mythical norm, which each one of us within our hearts knows "that is not me." In America, this norm is usually defined as white, thin, male, young, heterosexual, Christian, and financially secure. It is with this mythical norm that the trappings of power reside within this society. Those of us who stand outside that power often identify one way in which we are different, and we assume that to be the primary cause of all oppression, forgetting other distortions around difference, some of which we ourselves may be practicing.

Kahneman and Miller's (1986) theory addresses such normativity by arguing that category norm representations in working memory render the most common features of exemplars implicit. Their theory inspired studies examining how people spontaneously construct explanations for empirical group differences by taking higher status groups (such as men, heterosexuals, or White people) as the background norm for comparison and lower status groups (such as women, lesbians/gay men, and Black people) as the 'effect to be explained.' Later experimental research further demonstrated that such asymmetric explanations communicate the relative agency, power, status, and self-worth of those groups (Hegarty & Bruckmüller, 2013). Recently, Thorne (2018) examined the conceptualization of love between couples that vary by their partners' genders. Heterosexual participants' concepts of heterosexual love were closer to their default concept of love than was their concept of love between two women or two men. Sexual minority participants considered love between men but not love between women closer to their default, demonstrating a *homonormative* pattern, suggesting that their concepts of love had become more inclusive along lines of sexuality, but remained practically exclusive along lines of gender (see also Hegarty, Sczerba, & Skelton, 2019).

In the mid-1980s, it also started to become clear that similarity, could not explain why some things seemed to belong to the same category. Rather theories—networks of causal and explanatory links—act as the conceptual 'glue' to hold conceptual structures together (Murphy & Medin, 1985), and central features of a category are held together through conceptual theories combining context, function, and prior knowledge (e.g., Kempton, 1981; Medin, Lynch, Coley, & Atran, 1997; Medin & Ortony, 1989; Murphy & Wright, 1984). In the 1990s, theorists grew more interested in one explanation-based form of theory; *essentialism*—the assumption that categories have an underlying essence that causes and explains their diverse observable features (Medin & Ortony, 1989). Around this same time attempts were increasingly made to isolate and contain a theory of concepts that does not draw upon broader cultural knowledge reached its limit, and several researchers started to evaluate theories of concepts more extensively in terms of how these theories may make sense to ordinary people (e.g., Kelley, 1992; Komatsu, 1992; MacLaury, 1991). MacLaury (1991) argued that people may use more than one cognitive system to represent the same category. For example, the oddness and evenness of numbers can be categorized both classically (i.e., as categories with necessary and sufficient features) and as a category with graded structure (i.e., as a category with central and peripheral features) (Armstrong, Gleitman, & Gleitman, 1983), and the concept of 'doctor' can have both strict rules for category membership and be structured around a prototype (Dahlgren, 1985). This argument is post-positivist, diffractive and queer. It allows for the possibility of meaningful experimental phenomena, but does so by assuming not a static notion of categories, but one where categorization occurs in replicable interactions with scientific observers. As such people can show an *illusion of explanatory depth*; they think they understand the logical structure of their category structure more than they actually do (Hampton, 2010). Experimental research on category norms and explanations show that this illusion is highly asymmetric. As a result, such experimental research justifies the claims of critical psychologists that it is heterosexist to explain differences by focusing on attributes of nonheterosexual individuals and couples while conflating actual heterosexual individuals or couples with the default, ideal or norm (Herek, Kimmel, Amaro, & Melton, 1991), just as it is androcentric to reason about gender asymmetrically (Bailey, La France, & Dovidio, 2018) or a cultural misattribution bias (Causadias, Vitriol, & Atkin, 2018) to do so about race, ethnicity or culture.

Around the same time, queer theory started to emerge, in part from the ruins of a debate in lesbian and gay studies about whether it was more strategic to posit a discrete and immutable homosexual identity to account for very widespread historical and cultural evidence of homosexual acts, or if it were better to understand homosexuality as something 'socially constructed' by historical and cultural contexts (Stein, 1992). These conversations were sometimes described as essentialist-constructivist debates. Consistent with MacLaury's (1991) view, early queer theory texts argued that the essentialist-constructivist debates were irresolvable in absolute political or ethical terms; queer scholarship opposed an internally contradictory ideology that would construct new logical grounds to justify itself if it were threatened (Sedgwick, 1990). Queer theorists and cognitive psychologists independently came to the same conclusion, that human categorization is grounded in larger theories, and is not simply a reflection of stable or classical essences. Dominant cognitive miser theories in social cognition did not do justice to either development, and criticisms of social cognition as a cognitive miser theory may have missed what was most interesting and queer in the cognitive literature at this point in time.

Before moving to argue for the relevance of this intellectual history for contemporary intersex studies, we should note that cognitive psychologists often continued to exemplify their ideas in ways that naturalized essentialist understandings of sexuality and gender rather than took then as ideal objects of empirical inquiry. In an insightful synthesis of categorization research, Medin (1989, pp. 1476–1477) used gender attribution to exemplify psychological essentialism as follows.

> People in our culture believe that the categories male and female are genetically determined, but to pick someone out as male or female we rely on characteristics such as hair length, height, facial hair and clothing that represent a mix of secondary sexual characteristics and cultural conventions. Although these characteristics are more unreliable than genetic evidence, they are far from arbitrary. Not only do they have some validity in the statistical sense, but they are tied to our biological and cultural conceptions of male and female.

This quote closes our argument to diffract the findings from the ontology in cognitive psychology. Medin (1989) did not cite Kessler and McKenna (1978), but recognizes that we often categorize others on the basis of their inferred unobserved physical attributes, which are given a projected 'essential' explanatory status. Yet, Medin's quote enacts a dubious genetic essentialism of its own, constructing genes as the most reliable and statistically valid indicator of gender category membership, and tying them to a consensually shared biological and cultural two-gender system. As we show next, agency in intersex studies can be enabled by the ways in which categorization research psychologizes the very ordinariness of thinking queerly, but also be drawing on the insights of Kessler and McKenna (1978) which the best cognitive research on this topic—here and elsewhere—visibly misses.

## Implications for a Critical Psychology of Science and Technology

So far in this chapter, we have made the argument that people categorize the world in several different ways; this is evident in the cognitive psychology literature. More critically, people may use these different categorization systems flexibly depending on the context. How does this matter to the reach of a critical psychology of science and technology which this edited volume aims to extend? To address this question and to exemplify how categorizing queerly can be very ordinary, we turn our attention to the study of intersex.

The history of psychologists' affirmative engagements with intersex has, at this point in time, yet to be written (but see Hegarty, 2017, pp. 96–99). Psychology's problematic investment in intersex became more profound with the decades-long unprincipled practice initiated by psychologist John Money. Money developed his gender theory as the cognitive revolution was taking shape, and he took the children's natural cognitive capacity to learn any language as an analogy to undergird his claim that a child could learn any gender within a critical period (Morland, 2015). These assumptions undergirded recommendations for early 'corrective' surgeries on infants which are now a matter of global human rights concern, on the grounds that they enact bodily harm and deny rights to self-determination (Carpenter, 2016). Money's misreading of the cybernetic theory of his time and these later abuses are related; Money misunderstood how open the child's emerging gender—like any cybernetic system—could be to feedback and change, obfuscating the harm that is done by surgical interventions that aim to insert the child into the two-gender cultural system (Kessler, 1990; Morland, 2015). In other words, the historical entanglement of people with intersex traits and psychology—and the ethical questions about self-determination in the present—matters to the history of cybernetics, arguments about the uniquely human creativity expressed in language, the capacities of humans to use signify realities in multiple in commensurate systems, and the relationships between experts and laypeople's shared and contested understandings of the terms that are used to understand reality. The politics of neologism continue.

Since 2006, medical consensus has strongly proscribed the use of the term 'disorders of sex development' or DSD. We do not have space here to review the events that lead to this 'Chicago Consensus Statement' (but see Davis, 2015). The argument for DSD reproduced the ideal that good categories were classical, grounded in natural facts—such as genes, while bad categories had pejorative and controversial meanings. As a conceptual and linguistic intervention, DSD has been a practical success but an ontological failure with predictable negative consequences. Within medical debates, experts disagree as to whether genetic features or genital anatomy constitute the essential features of the DSD category, with competing professional interests on both sides of the debate, and the inclusion or exclusion of common forms of embodiment such as Turner's syndrome and Klinefelter's syndrome hanging in the balance (Griffiths, 2018).

However, recent psychosocial research shows how the invention of DSD recapitulates the errors of assuming categories to be good only if they are classical categories with clearly defined boundaries. Diagnostic categories delivered by clinicians fail to provide the conceptual flexibility demanded by everyday life (Lundberg, Linstrom, Roen, & Hegarty, 2016). In practice many people need to alter the volume on these categories, and to compartmentalize them rather than assume that they are all-defining (Lundberg, Roen, Hirschberg, & Frisen, 2016). For example, young people can also combine the language of 'intersex' and 'DSD' with fluency. As one young person with an intersex variation in Lundberg, Hegarty, and Roen's (2018, p. 167) study put it: 'I think DSD just describes physically how my sex development has been different and Intersex just describes how I feel like my gender identity is maybe not a 100% female.' Young people with intersex traits are variably aware of DSD and intersex as categories that others have used to name their experience. They and their parents, in different contexts, avoid using them, use them to describe traits but not people, use them interchangeably, and adopt them as social identities. Such flexibility is necessary in lifeworlds in which people must orient to managing privacy under the threat of others' fascination, form meaningful identities and relationships, and communicate in medicalized and nonmedical environments about their wishes and needs while keeping options for revision open down the line (Lundberg et al., 2018). The ordinariness of this is paramount to our argument about how natural and ordinary it is for individuals to apply different categorization systems to understand their own social identities—a far cry from the image of the 'cognitive miser.' It would be miserly to insist upon a more rigid, fixed, consistent, or trans-situational approach to making sense of oneself or one's children. It is hard to escape the conclusion that such attempts do indeed re-extend the reach of the scientific and medical authority into the lived realities of such individuals and their families.

It is simple and ordinary for people to switch between categories, and between forms of representing those categories via exemplars, traits, or rules of definition in discourse. Social identity theorists have long explored these questions using both discursive and experimental frameworks of understanding, often in communication with each other. Working in that tradition, Morgenroth and Ryan (2018) have recently argued for a return to Judith Butler's work in social psychology to ground understandings of the flexibility and performativity of gender, as in Deaux and Major's (1987) model of gender stereotyping, for example. Intersex takes us beyond gender to showing the material consequences of thinking about category complexity as ordinary in regard to the impact that variable sex characteristics can make to lives lived in socio-technical societies. As Lundberg et al.'s (2018) work suggests, language terms do not simply signify anatomic features here but *theories* about the ways in which bodies and selves are or are not related. Those theories communicated by language use afford ground for materially different medical decisions with lifelong consequences (e.g., Streuli, Vayena, Cavicchia-Balmer, & Huber, 2013). By critically engaging in post-positivist and queer thinking *about thinking*, critical psychology can engage these dynamics and make difference in science and reality.

## Conclusion

In conclusion, we argued here for a diffractive reading of categorization research. To be sure, the logic of cognitive psychology itself needs to be read critically, as our focus on the validity of the genetic basis of gender was meant to exemplify above. However, ours is not an idealist argument against fuzzy meaning per se; quite the opposite. Nothing as abstract as categorization—or sexuality, or desire or identification—can be communicated, or become a basis for socially shared understanding without structuring metaphors and exemplars, and we consider communication that aims at such shared understandings about these features of human experience to be a worthy and worthwhile thing for psychologists concerned with sense-making to do. As such we read cognitive research diffractively (Barad, 2007), for its replicable demonstrations of queer thinking that emerge in the research interactions we call experiments (see also Scholz, 2013, for an application of Barad's thought to experimental psychology), and for the ways that its discourse has guided understanding of what minds might be.

Engagement with psychology's mainstream in lesbian and gay affirmative psychology has often been called a form of 'strategic essentialism' (e.g., Kitzinger & Coyle, 2000). We hope that this diffractive reading offers critical psychologies of science and technology a supplementary exemplar of *strategic anti-essentialist* reading. Over the long run since the early cognitive revolution, the Popperian logic of cognitive psychology—which values falsifying preexisting theories with experimental findings—has created an opportunity structure for experiments that attempt to falsify prior theories for their simplifications now for several decades. The emphasis on falsification over replication has more recently become a matter of considerable debate *within* experimental psychology, making it timely to remember that Popper (1959/2002, p. 37) defined falsification, the criterion for demarcating science from non-science, in explicitly *social* and *conventional* terms. For Popper, the suitability of any criterion of falsifiability was not given by logic or material reality, but might be a matter of debate and discussion that rests on matters other than reason, as Potter (1984) has observed it to be in psychologists' discourses.

In the era of the current replication crisis, reasons to reject experimental psychology's truth claims as erroneously socially constructed have never been more available, nor more obviously embraced. Our reading goes beyond reflection on how all-too human research practices are ironically at odds with the ontologies of mind that they produce, to argue that taking the results of such experiments seriously might be a form of enabling agency. Rather than doubt cognitive psychology we argue for an analytic way through its empirical literature because different readings of cognitive psychology change the interpretation of human action in the world. We made this case by looking at how contemporary linguistic politics in intersex are read with two different theories of the individual subject are held in mind; in which gender and linguistic capacity for flexibility are queer, in the pejorative sense and in the sense that thinking queerly is deemed ordinary by cognitive accounts of sense-making. We hope to have gone at least half-way in making these seemingly different worlds meet in the middle.

**Acknowledgements** We would like to thank Jeffrey Yen, Kieran O'Doherty, David Griffiths, and Jacy Young for their readings and comments on our initial draft of this chapter.

## References

Armstrong, S. L., Gleitman, L. R., & Gleitman, H. (1983). What some concepts might not be. *Cognition, 13,* 263–308.

Bailey, A. H., LaFrance, M., & Dovidio, J. F. (2018). Is man the measure of all things? A social cognitive account of androcentrism. *Personality and Social Psychology Review.*

Barad, K. (2007). *Meeting the universe half-way: Quantum physics and the entanglement of matter and meaning.* Durham, NC: Duke University Press.

Barsalou, L. W. (1982). Context-independent and context-dependent information in concepts. *Memory & Cognition, 10,* 82–93.

Barsalou, L. W. (1983). Ad hoc categories. *Memory & Cognition, 11,* 211–227.

Berlin, B., & Kay, P. (1969). *Basic color terms: Their universality and evolution.* Berkeley: University of California Press.

Bertolini, V. J. (1996). Fireside chastity: The erotics of sentimental bachelorhood in the 1850s. *American Literature, 68,* 707–737.

Bowker, G. C., & Starr, S. L. (1999). *Sorting things out: Classification and its consequences.* Cambridge: MIT Press.

Brown, L. S. (1989). New voices, new visions: Toward a lesbian/gay paradigm for psychology. *Psychology of Women Quarterly, 13,* 445–458.

Bruner, J. (1990). *Acts of meaning.* Cambridge, MA: Harvard University Press.

Butler, J. (1990). *Gender trouble and the subversion of identity.* New York and London: Routledge.

Carpenter, M. (2016). The human rights of intersex people: Addressing harmful practices and rhetoric of change. *Reproductive Health Matters, 24,* 74–84.

Causadias, J. M., Vitriol, J. A., & Atkin, A. L. (2018). Do we overemphasize the role of culture in the behavior of racial/ethnic minorities? Evidence of a cultural (mis)attribution bias in American psychology. *American Psychologist, 73,* 243–255.

Clarke, V., Ellis, S. J., Peel, E., & Riggs, D. W. (2010). *Lesbian, gay, bisexual, trans and queer psychology: An introduction*. Cambridge: Cambridge University Press.
Dahlgren, K. (1985). The cognitive structure of social categories. *Cognitive Science, 9,* 379–398.
Davis, G. (2015). *Contesting intersex: The dubious diagnosis*. New York: New York University Press.
Deaux, K., & Major, B. (1987). Putting gender into context: An interactive model of gender-related behavior. *Psychological Review, 94,* 369–389.
DeLamater, J. D., & Hyde, J. S. (1998). Essentialism vs. social constructionism in the study of human sexuality. *Journal of Sex Research, 35,* 10–18.
Diamond, L. M. (2003). What does sexual orientation orient? A biobehavioral model distinguishing romantic love and sexual desire. *Psychological Review, 110,* 173–192.
Diamond, L. M. (2004). Emerging perspectives on distinctions between romantic love and sexual desire. *Current Directions in Psychological Science, 13,* 116–119.
Downing, L., & Gillett, R. (2011). Viewing critical psychology through the lens of queer. *Psychology & Sexuality, 2,* 4–15.
Dunham, Y., & Olson, K. R. (2016). Beyond discrete categories: Studying multiracial, intersex, and transgender children will strengthen basic developmental science. *Journal of Cognition and Development, 17,* 642–665.
Fallon, A. E., & Rozin, P. (1985). Sex differences in perceptions of desirable body shape. *Journal of Abnormal Psychology, 94,* 102–105.
Foucault, M. (1975). *Discipline and punish: The birth of the prison*. New York: Random House.
Foucault, M. (1998). *The history of sexuality: 1*. London: Penguin.
Genero, N., & Cantor, N. (1987). Exemplar prototypes and clinical diagnosis: Toward a cognitive economy. *Journal of Social and Clinical Psychology, 5,* 59–78.
Griffiths, D. A. (2018). Diagnosing sex: Intersex surgery and 'sex change' in Britain 1930–1955. *Sexualities, 21,* 476–495.
Halperin, D. M. (1995). *Saint Foucault: Towards a gay hagiography*. New York: Oxford University Press.
Hammack, P. L., & Cohler, B. J. (2009). *The story of sexual identity: Narrative perspectives on the gay and lesbian life course*. New York: Oxford University Press.
Hampton, J. A. (1987). Inheritance of attributes in natural concept conjunctions. *Memory & Cognition, 15,* 55–71.

Hampton, J. A. (2010). Concepts in human adults. In D. Mareschal, P. Quinn, & S. E. G. Lea (Eds.), *The making of human concepts* (pp. 293–311). Oxford: Oxford University Press.

Hastie, R., Schroeder, C., & Weber, R. (1990). Creating complex social conjunction categories from simple categories. *Bulletin of the Psychonomic Society, 28,* 242–247.

Hegarty, P. (2001). "Real science", deception experiments and the gender of my lab coat: Toward a new laboratory manual for lesbian and gay psychology. *Critical Psychology, 1,* 91–108.

Hegarty, P. (2002). 'It's not a choice, it's the way we're built': Symbolic beliefs about sexual orientation in the US and Britain. *Journal of Community & Applied Social Psychology, 12,* 153–166.

Hegarty, P. (2007). From genius inverts to gendered intelligence: Lewis Terman and the power of the norm. *History of Psychology, 10,* 132–155.

Hegarty, P. (2017). *A recent history of lesbian and gay psychology: From homophobia to LGBT.* London: Routledge.

Hegarty, P., & Bruckmüller, S. (2013). Asymmetric explanations of group differences: Experimental evidence of Foucault's disciplinary power. *Social and Personality Psychology Compass, 7,* 176–186.

Hegarty, P., Sczerba, A., & Skelton, R. (2019). How has cultural heterosexism affected thinking about divorce? Asymmetric framing of same-gender and mixed-gender divorces in news media and in minds. *Journal of Homosexuality.*

Herek, G. M., Kimmel, D. C., Amaro, H., & Melton, G. B. (1991). Avoiding heterosexist bias in psychological research. *American Psychologist, 46,* 957–963.

Heyes, C. J. (2006). Foucault goes to weight watchers. *Hypatia, 21,* 126–149.

Hyde, J. S., Bigler, R. S., Joel, D., Tate, C. C., & van Anders, S. M. (2018). The future of sex and gender in psychology: Five challenges to the gender binary. *American Psychologist, 74,* 171–193.

Jagose, A. (1996). *Queer theory: An introduction.* New York: New York University Press.

Johnson, K. (2015). *Sexuality: A psychosocial manifesto.* Cambridge: Polity Press.

Kahneman, D., & Miller, D. T. (1986). Norm theory: Comparing reality to its alternatives. *Psychological Review, 93,* 136–153.

Kelley, H. H. (1992). Common-sense psychology and scientific psychology. *Annual Review of Psychology, 43,* 1–23.

Kempton, W. (1981). *The folk classification of ceramics: A study of cognitive prototypes*. New York: Academic Press.

Kessler, S. J. (1990). The medical construction of gender: Case management of intersexed infants. *Signs: Journal of Women in Culture and Society, 16,* 3–26.

Kessler, S. J., & McKenna, W. (1978). *Gender: An ethnomethodological approach*. New York: Wiley.

Kitzinger, C. (1997). Lesbian and gay psychology: A critical analysis. In D. Fox & I. Prilleltensky (Eds.), *Critical psychology: An introduction* (pp. 202–216). Thousand Oaks: Sage.

Kitzinger, C., & Coyle, A. (2000). Introducing lesbian and gay psychology. In C. Kitzinger & A. Coyle (Eds.), *Lesbian and gay psychology: New perspectives*. Leicester, UK: BPS Blackwell.

Komatsu, L. K. (1992). Recent views of conceptual structure. *Psychological Bulletin, 112,* 500–526.

Kunda, Z., Miller, D. T., & Claire, T. (1990). Combining social concepts: The role of causal reasoning. *Cognitive Science, 14,* 551–577.

Kunda, Z., & Oleson, K. C. (1995). Maintaining stereotypes in the face of disconfirmation: Constructing grounds for subtyping deviants. *Journal of Personality and Social Psychology, 68,* 565–579.

Lorde, A. (1984). *Sister outsider: Essays and speeches*. Berkeley, CA: Crossing Press.

Lundberg, T., Hegarty, P., & Roen, K. (2018). Making sense of 'Intersex' and 'DSD': How laypeople understand and use terminology. *Psychology & Sexuality, 9,* 161–173.

Lundberg, T., Lindström, A., Roen, K., & Hegarty, P. (2016). From knowing nothing to knowing what, how and now: Parents' experiences of caring for their children with congenital adrenal hyperplasia. *Journal of Pediatric Psychology, 42,* 520–529.

Lundberg, T., Roen, K., Hirschberg, A. L., & Frisén, L. (2016). "It's part of me, not all of me": Young women's experiences of receiving a diagnosis related to diverse sex development. *Journal of Pediatric and Adolescent Gynecology, 29,* 338–343.

Machery, E. (2009). *Doing without concepts*. New York: Oxford University Press.

MacLaury, R. E. (1991). Prototypes revisited. *Annual Review of Anthropology, 20,* 55–74.

Margolis, E. (1994). A reassessment of the shift from the classical theory of concepts to prototype theory. *Cognition, 51,* 73–89.

Medin, D. L. (1989). Concepts and conceptual structure. *American Psychologist, 44,* 1469–1481.
Medin, D. L., Lynch, E. B., Coley, J. D., & Atran, S. (1997). Categorization and reasoning among tree experts: Do all roads lead to Rome? *Cognitive Psychology, 32,* 49–96.
Medin, D. L., & Ortony, A. (1989). Psychological essentialism. In S. Vosniadou & A. Ortony (Eds.), *Similarity and analogical reasoning* (pp. 179–196). Cambridge: Cambridge University Press.
Medin, D. L., Ross, N. O., Atran, S., Cox, D., Coley, J. D., Proffitt, J. B., & Blok, S. (2006). Folkbiology of freshwater fish. *Cognition, 99,* 237–273.
Medin, D. L., & Schaffer, M. M. (1978). Context theory of classification learning. *Psychological Review, 85,* 207–238.
Mervis, C. B., & Rosch, E. (1981). Categorization of natural objects. *Annual Review of Psychology, 32,* 89–115.
Morgenroth, T., & Ryan, M. K. (2018). Gender trouble in social psychology: How can Butler's work inform experimental social psychologists' conceptualization of gender? *Frontiers in Psychology, 9,* 1320.
Morland, I. (2015). Gender, genitals and the meaning of being human. In L. Downing, I. Morland, & M. Sullivan (Eds.), *Fuckology: Critical essays on John Money's diagnostic concepts* (pp. 69–98). Chicago, IL: University of Chicago Press.
Murphy, G. L. (1988). Comprehending complex concepts. *Cognitive Science, 12,* 529–562.
Murphy, G. L., & Medin, D. L. (1985). The role of theories in conceptual coherence. *Psychological Review, 92,* 289–316.
Murphy, G. L., & Wright, J. C. (1984). Changes in conceptual structure with expertise: Differences between real-world experts and novices. *Journal of Experimental Psychology: Learning, Memory, and Cognition, 10,* 144–155.
Popper, K. R. (2002). *The logic of scientific discovery*. London: Routledge. (Original work published 1959).
Potter, J. (1984). Testability, flexibility: Kuhnian values in scientists' discourse concerning theory choice. *Philosophy of the Social Sciences, 14,* 303–330.
Rivers, I. (2000). Counting matters: Quantitative research in lesbian and gay psychology. *Lesbian & Gay Psychology Review, 1,* 28–31.
Rosch, E. (1973). On the internal structure of perceptual and semantic categories. In T. E. Moore (Ed.), *Cognitive development and the acquisition of language* (pp. 111–144). New York: Academic Press.
Rosch, E. (1975). Cognitive representations of semantic categories. *Journal of Experimental Psychology: General, 104,* 192–233.

Rosch, E., & Mervis, C. B. (1975). Family resemblances: Studies in the internal structure of categories. *Cognitive Psychology, 7,* 573–605.

Rosch, E., Simpson, C., & Miller, R. S. (1976). Structural bases of typicality effects. *Journal of Experimental Psychology: Human Perception and Performance, 2,* 491–502.

Roth, E. M., & Schoben, E. J. (1983). The effect of context on the structure of categories. *Cognitive Psychology, 15,* 346–378.

Scholz, J. (2013). The possibility of a quantitative queer psychology. In K. O'Mara & L. Moorish (Eds.), *Queering paradigms III: Queer impact and practices* (pp. 239–258). Oxford: Peter Lang.

Sedgwick, E. K. (1990). *Epistemology of the closet.* Berkeley, CA: University of California Press.

Smith, E. E., & Medin, D. L. (1981). *Categories and concepts.* Cambridge: Harvard University Press.

Smith, E. E., & Osherson, D. N. (1984). Conceptual combination with prototype concepts. *Cognitive Science, 8,* 337–361.

Smith, E. E., Osherson, D. N., Rips, L. J., & Keane, M. (1988). Combining prototypes: A selective modification model. *Cognitive Science, 12,* 485–527.

Solomon, K. O., Medin, D. L., & Lynch, E. (1999). Concepts do more than categorize. *Trends in Cognitive Sciences, 3,* 99–105.

Stein, E. (Ed.). (1992). *Forms of desire: Sexual orientation and the social constructionist controversy.* New York: Routledge.

Storms, G., De Boeck, P., Van Mechelen, I., & Ruts, W. (1996). The dominance effect in concept conjunctions: Generality and interaction aspects. *Journal of Experimental Psychology. Learning, Memory, and Cognition, 22,* 1–15.

Streuli, J. C., Vayena, E., Cavicchia-Balmer, Y., & Huber, J. (2013). Shaping parents: Impact of contrasting professional counseling on parents' decision making for children with disorders of sex development. *Journal of Sexual Medicine, 10,* 1953–1960.

Tajfel, H., & Turner, J. C. (1986). The social identity theory of intergroup behaviour. In S. Worchel & W. G. Austin (Eds.), *Psychology of intergroup relations.* Chicago: Nelson-Hall.

Taylor, S. E., Fiske, S. T., Etcoff, N. L., & Ruderman, A. J. (1978). Categorical and contextual bases of person memory and stereotyping. *Journal of Personality and Social Psychology, 36,* 778–793.

Teo, T. (2010). What is epistemological violence in the empirical social sciences? *Social and Personality Psychology Compass, 4,* 295–303.

Thorne, S. (2018). *Queer concepts of romantic love: Uncovering a heteronormative bias*. Unpublished Ph.D. dissertation, University of Surrey, UK.
Tolman, E. C. (1948). Cognitive maps in rats and men. *Psychological Review, 55,* 189–208.
Turner, W. B. (2000). *A genealogy of queer theory*. Philadelphia: Temple University Press.
Warner, D. N. (2004). Towards a queer research methodology. *Qualitative Research in Psychology, 1,* 321–337.
Wason, P. C. (1960). On the failure to eliminate hypotheses in a conceptual task. *Quarterly Journal of Experimental Psychology, 12,* 129–140.
Wittgenstein, L. (1953). *Philosophical investigations*. New York: Macmillan.

# 13

# A Gendered Prestige: The Powers at Play When Doing Psychology with Ink Blots/ Statistics

**Katherine Hubbard and Natasha Bharj**

In this chapter, we outline the perhaps surprising commonalities between women and the Rorschach ink blot test (Rorschach, 1921) in the history of Psychology. Both have been considered 'subjective,' easily influenced, and of having the opposite attributes required for 'objective' science. Statistics, by contrast, have been conceptualized in a more masculine manner—that is: objective, logical, and resistant to influence. In drawing such comparisons, it is not our intention to argue for or against the Rorschach's reliability, validity, or indeed whether it *works*. Nor, are we making claims about the reliability or validity of statistical procedures in Psychology. Such a debate is not within our interests. Instead, we

K. Hubbard (✉)
Department of Sociology, University of Surrey,
Guildford, Surrey, UK
e-mail: k.a.hubbard@surrey.ac.uk

N. Bharj
Department of Psychology, University of Kansas,
Lawrence, KS, USA
e-mail: n527b291@ku.edu



K. C. O'Doherty et al. (eds.), *Psychological Studies of Science and Technology*, Palgrave Studies in the Theory and History of Psychology,
https://doi.org/10.1007/978-3-030-25308-0_13

are interested in how constructions of 'science' are implicated by ideas about gender and different tools used by Psychologists. We do not argue that either the Rorschach or statistical methods are legitimate or objective, but rather Psychology incorporates subjectivity in *all* areas and that wider social beliefs structure decisions about what is and is not considered legitimate. Specifically, we are drawing connections between beliefs about women and beliefs about projective tests (such as the Rorschach), and contrasting these to the history of belief in statistics in Psychology. In doing so, we provide a short analysis of the epistemic powers at play in the history of Psychology and its construction as a legitimate science.

Both the Rorschach and statistics have, at various times, been powerful tools in the hands of psychologists. Constructions of these tools as valid (or not) are revelatory of the discursive powers at work when psychologists decide what is legitimate, objective and scientific. By treating the discipline of Psychology as our subject matter (see Richards, 2002), we align ourselves with one of the aims of this book, to view Psychology through a lens of Science and Technology Studies. Specifically, in drawing together interdisciplinary thought, including feminist approaches, we offer some demonstration as to how societal beliefs impact the technologies and tools utilized by psychologists and vice versa. Entrenched within such explorations of gender, scientific legitimacy, and construction of knowledge, is power. Therefore, this chapter will centrally consider the power dynamics working within these histories and how such power contributed to the historically gendered nature of prestige within Psychology.

In the following, we first consider how polarizations of subjective/objective, women/men, invalid/valid are entrenched within ideas of what science ought to look like. This emerges in the next section as important for not only what the *science* looks like but also what the scien*tist* looks like. Second, we briefly outline the history of women in Psychology with particular reference to those involved in projective tests and the Rorschach. Finally, we consider one case study where the Rorschach, a woman psychologist, and the use of statistics, came together. In this example, we hope to show how belief in psychological methods and tests is key to viewing Psychology as legitimate, rather than an inherent legitimacy or objectivity. As illustrated in this example, such beliefs have real social consequences for marginalized groups.

## (God-Trick [Prestige ~ Power])

In 2000 the APA Task Force on Women in Academe reported on the obstacles and inequities that prevent women from fully participating in research and leadership. These included pressures for women to conform to gender stereotypes by over-performing service, such as having heavy loads of committee administrative work and mentoring. Studies across academia, especially STEM, show women continue to be dis-associated from traits valued within positivist epistemologies, such as agency and scientific competency (Madera, Hebl, & Martin, 2009; Rees, 2011). These processes, not to mention wage disparities and the ever-present reality of sexual harassment, perpetuate the 'leaky pipeline' of women's career development in academia (Gasser & Shaffer, 2014).

Beyond such structural barriers, women in Psychology have had to contend with their politicized presence in a field that eschews explicit politics. Naomi Weisstein challenged the supposed ideological neutrality of psychological research by endorsing the use of experimental methods as a political tool, and traced the importance of social context in the activity of nerve cells (Rutherford, Vaughn-Blount, & Ball, 2010). Feminist psychologists have made great strides in dispelling the gender essentialism and prejudices that maintained this tension in the past (see Hyde, 1990), yet as demonstrated in the brief examples above, the devaluing and (pejorative) politicizing of women's psychological work remains. Gendered epistemological power, as well as lingering prejudice, structures this imbalance of prestige.

Psychology, in its hegemonic Western form, rests upon a foundation of positivist epistemology and a scientific method borrowed from the natural sciences (Gergen, 1973). Aspirations of objectivity hold the greatest prestige as the key to 'pure' inquiry. At one time in Psychology, this was thought to be best achieved through the Rorschach ink blots test (especially in the US), now it is embodied by statistical methods. Similarly, it was also thought to be best embodied by men as achieving the most objective science (Madera et al., 2009; Rees, 2011). Hegarty (2007, p. 83) described the notion of scientific 'purity' as inherently embroiled with power and calls for:

> psychologists, and historians of psychology, to collectivity consider how our unthinking attempts at objectivity, impartiality, and expertise might be motivated by anxieties about positioning ourselves on the safe side of hierarchal value-laden category boundaries which go unspoken.

These boundaries (objective/subjective, nature/nurture, hard/soft science) position un marked-ness and outsider status as necessary to practice scientific inquiry—performing Haraway's "God trick … promising vision from everywhere and nowhere equally and fully" (1988, p. 584). The centrality of the god-trick to positivist scientific practice draws sharp boundaries between those whose view is marked or unmarked. The supposition that women's perspectives are inherently subjective, determined by internal processes of emotion and physicality, locates women as a group firmly *somewhere*—outside of the 'nowhere' required for objective observation. Men, who are not marked as 'having' gender, can therefore occupy an objective and dislocated position from which to construct universal knowledge. Decolonial theories identify the same processes in the "zero point epistemology" used to abstract Western, colonial perspectives from context; Western scientific thought became hegemonic through constructing racially and ethnically marked (colonized) people as inherently geographically, historically, and physically grounded, and therefore incapable of the universal and objective thought produced by racially unmarked White Europeans (Mignolo, 2011; p. 80). Being 'marked' has historically positioned that person as further from the white male norm with which concepts of 'objectivity' are aligned. This symbolic asymmetry developed throughout the history of Western science; the gendering of mind/body dualism (men as mind, women as body) during the Enlightenment proliferated to contemporary gender perceptions that mark women as representatives of their groups, while men remain individuals (Amâncio & Oliveira, 2006). The same can be said of the marking of people of color as representatives of their culture, while white people remain racially and culturally unmarked (Causadias, Vitriol, & Atkin, 2018). Like Haraway (1988), we are critical of this idea of objectivity—that scientists are able to abstract themselves from their object of study, or that it would even be desirable to do so.

Feminist Psychology has responded to women's exclusion from the god-trick of positivist science using two major strategies; reframing positivism to include feminist epistemologies, and challenging the devaluing of situated perspectives. Feminist standpoint theory emphasizes the epistemic privilege of the 'view from below'; oppressed groups are better able to see and articulate sociopolitical structures by virtue of being subject to them (Haraway, 1988; Wylie, 2004). This approach pushes back against the devaluing of subjectivity that has left women at the margins of Psychology. Feminist philosophers of science have attempted to reposition feminist epistemologies within positivist frameworks by redefining notions of objectivity to allow for epistemic advantage to be recognized (Harding, 1992). Despite this, feminist work, particularly feminist qualitative research, continues to be disparaged as overly influenced by the personal and political, in opposition to statistics and 'unmarked' work conducted by 'unmarked' researchers.

As psychological tools, the practices of quantitative analyses and experimental designs involve politicized and subjective interactions between researcher-and-materials and researcher-and-subject. Yet, statistics and experimental methods have been taken up as the sole tools of positivist epistemology, and stripped of their subjectivity and politics. The politics of statistics and the failure of conventional experimental procedures to meet standards of objectivity are under-articulated, allowing for their prestige to be maintained (see Spears & Smith, 2001). The recent replication crisis in Psychology signals a potential paradigm shift; anxieties over the influence of subjectivity on statistical analyses have led to calls for strict rules of practice (e.g., preregistration) and analysis (e.g., requirements for more extensive reporting of analyses) (Rovenpor & Gonzales, 2015). What has failed to culminate from this 'crisis' is an open dialogue on the epistemological assumptions underlying statistical practice and whether alternative epistemic models might benefit the future of psychological work.

## Gender, the Rorschach, and Statistics

As evidenced in the above, socially entrenched ideas about gender have been enduring, and Victorian and post-Darwinian conceptualizations were clearly evident in the early stages of Psychology becoming an organized discipline. Shields (2007) has argued that women's traits were seen as being naturally complementary and inferior to men's traits. For example, she quotes Victorian Psychiatrist Henry Maudsley who said in 1879 "the life is more developed in proportion to the intellect in the female than in the male, and affective the influence of the reproductive organs upon mind more powerful." A few decades later, Maudsley gave (initially anonymously) a substantial sum of money to London County Council to open a new psychiatric Hospital which finally opened in 1923. The 'Maudsley Hospital' became one of the main mental health hospitals and sites for Clinical Psychology training in Britain. It was at the Maudsley that the wave of criticism toward projective tests in Britain first emerged. For example, one-time keen eugenicist Aubrey Lewis was Chair of Psychiatry at the Institute of Psychiatry positioned at the Maudsley in 1946 and had described the Rorschach ink blot test as of 'limited or doubtful value' as early as 1934. This one example illustrates how the development of Clinical Psychology in Britain was tied with (a) thinking about projective tests and (b) attitudes toward women.

In contrast to men's logic, mental strength and intelligence, Shields (2007) argued women were positioned as easily influenced, damaged and vulnerable. Women were framed as naturally submissive and inferior and so were not encouraged into education for the most part. Gender essentialism in part structured these framings; uteri were believed to be more likely to cause havoc on the mind and body than testes, and these supposed physiological differences in turn impacted societal beliefs about physical strength and the ability to be educated. Such beliefs about women's bodies and physical capabilities continued well into the twentieth century: it was not until 1984 that women were allowed by Olympic officials to run a marathon race following substantial protest and action by women athletes (indeed, it was not until 2008 that women's 3000 m steeplechase was included, see Burfoot, 2016).

That is not to say, however, that women were not present within the history of Psychology, even at the very beginning (or that they did not run marathons before 1984) (see Fancher & Rutherford, 2012). This is the key issue of how history gets told—those who are considered 'legitimate' and are in power, are often those who get to choose which stories get told, retold and *how* they get told. The efforts of women have therefore often gone unspoken within the *his*tory of Psychology

(Bernstein & Russo, 1974; Furumoto, 2003). Because of the misogynistic positioning of women throughout the majority of the twentieth century and androcentric history telling, the global effort to re-place women in the history of Psychology has been undertaken by feminist psychologists and historians.

Since the end of the twentieth century the feminist action to re-place women in the history of Psychology has gained substantial traction (see Bohan, 1990; Furumoto & Scarborough, 1986; Morawski & Agronick, 1991; Rutherford, Vaughn-Johnson, & Rodkey, 2015; Scarborough & Furumoto, 1989). The venture to write women's history in Psychology is ongoing and one of the major contemporary projects is Psychology's Feminist Voices.[1] In the US, Furumoto and Scarborough (1986) studied the lives of the first 22 women psychologists who achieved their doctorates around the turn of the twentieth century. All of those who attained assistant professorship or higher were unmarried and each experienced discrimination. Milar (2000) found that in the first group of women psychologists, only 50% had a professional rank compared to 65% of psychologists who were men. All of those professional women were single and worked predominantly in women's colleges; many also had to work for free or for very little pay. Most women colleges only employed unmarried women. Such 'marriage bans' did not take full effect in British Universities, though Liverpool University did try to establish one (Valentine, 2008).

Women, unlike men, were pressured to choose *either* career or marriage (Milar, 2000). Still, both Oxford and Cambridge Universities were reluctant to accept women students (Shields, 2007). Higher education in both the US and in Britain was often only available to higher class women who had independent finances or were supported by rich relatives- some women at this time were said to 'rebel' against their fathers in order to gain doctoral level educations (e.g., Margaret Lowenfeld, see Hubbard, 2018). Because women were less likely to be afforded

[1]See http://www.feministvoices.com/ a project which provides first-hand accounts of feminist psychologists and highlights women's contributions to Psychology's past and recognizes the voices of contemporary feminist psychologists.

opportunities for academic training and careers in Psychology, their work remained largely invisible and was less often cited (Stevens & Gardners, 1982).

Nevertheless, in Britain women were present at the beginning of Psychology, though their work often went largely unrecognized. Valentine (2008, 2010) considered the positions of women in early British Psychology, and suggested that women found Psychology as a discipline more accessible than other sciences, especially Physiology. This was perhaps, Valentine (2008) suggests, because of the efforts to increase the numbers of the recently formed British Psychological Society (BPS). Or perhaps, particularly after the First World War, because Psychology was such a new science that men's dominance had yet to gain a foothold. The majority of the women involved in early Psychology were middle/upper-class, worked in teaching roles, and two-thirds (11/16) were unmarried. However, despite women's presence, gender issues remained. For example, Alice Woods, one of the founding members of the BPS, described how in 1913 all of the women attending the very first reading of Sigmund Freud's work were asked to leave the room (Valentine, 2008).

Other areas of Psychology were similarly resistant to the presence of women. In the US the postwar 'Servicemen's Readjustment Act' (1944) and the 'Vocational Rehabilitation Act,' commonly known as the 'G.I. Bill' prioritized veterans who wished to train in Psychology, increasing the influx of men into Clinical Psychology. Many military positions for psychologists were closed to women entirely (Bohan, 1990). From the 1950s new members of the Committee of Professional Psychologists in Britain were increasingly likely to be men and Clinical Psychology as a subdiscipline began to dominate Psychology as responses to war-related trauma were needed (Hall, 2007). There was also a deliberate attempt following the war to not only provide men with jobs (including in Psychology), but also a keen desire to ensure women returned to their roles as homemakers again following the relative occupational opportunity afforded during the War (Morawski & Agronick, 1991).

Following the Second World War, there were some areas of Psychology that were considered better suited to women. Child, or developmental, Psychology was one such area, due to societal beliefs about women's apparently natural abilities in child-centered nurturing, nursing and care work (Furumoto & Scarborough, 1986; Rutherford et al., 2015; Scarborough & Furumoto, 1989). Stereotypes of women's nurturing abilities opened Developmental Psychology as the most appropriate field for women; and it was within this framework that Mamie Phipps Clark developed her Master's thesis on racial identification. The boundaries of science/politics and objectivity/subjectivity permeated this field still; when Clark, along with her husband Kenneth, applied this research to the *Brown vs Board of Education of Topeka* decision there was dissent among legal scholars who asserted the incompatibility of scientific research and politics (Guthrie, 1990).

Another area which was considered suitable for women was psychological testing. Furumoto (2003) argues that these testing practices had a large impact on the overall development of Psychology, especially in its unprecedented growth after the First World War. Testing was viewed, despite its importance and impact upon Psychology, as lower status and was thought to require less technical knowledge. It was therefore deemed suitable for women and so provided them with opportunities in lower salaried jobs than their men counterparts (Bohan, 1990). Unsurprisingly then, testing boomed in areas such as employment, educational and developmental Psychology where there were higher concentrations of women working (Bohan, 1990; Furumoto, 2003). Under the control of mainly women psychologists, applied Psychology and testing practices greatly advanced the profession of Psychology.

One area of testing which was particularly prominent following the Second World War was projective testing. Projective tests are those which provide ambiguous stimuli and the person being tested is said to 'project' their psychology onto their interpretation of the stimuli. The most famous projective test and the one which was most successful in terms of popularity was the Rorschach ink blot test (Rorschach, 1921). The Rorschach ink blot test became the most used psychological test following the Second World War in the US, having been used to test potential officers in the US military (see Hegarty, 2003a; Hubbard & Hegarty, 2016). The Rorschach was also used to 'detect' gay men and those malingering as gay in the Second World War in the US (Hegarty, 2003a). With the Rorschach at the center, projective testing grew in popularity all over the world including the US (Buchanan, 1997; Brunner, 2001; Hegarty, 2003a; Lemov, 2011); and Britain (Hubbard & Hegarty, 2016; McCarthy Woods, 2008).

Largely mirroring the patterns of popularity of projective tests in the US, Britain's projective test movement gained a foothold in the 1940s, resulting in a dedicated journal and society (Hubbard & Hegarty, 2016). In Britain especially, there was a higher proportion of women involved in the projective testing movement compared to other areas of Psychology. For example, in the British Rorschach Forum in 1958, there were eight women and five men on the committee, and women occupied 62–71% of committee positions until 1969. In December 1966, a register showed that 48% of all society fellows, members, and associates were women. Among the first authors of publications in *The Rorschach Newsletter* from 1952 to 1968, 41% were women. However, following the 1968 International Rorschach Congress which was held in London and headed up by Theodora Alcock, the presence of men increased. At the December 1968 Annual General Meeting, just months after the International Rorschach Congress, one woman and seven men were elected onto the committee of the re-named British Rorschach Forum and Society for Projective Techniques (see Hubbard & Hegarty, 2016).

Therefore, women have been present in the history of Psychology, but the areas in which they were able to negotiate access and practice were those areas deemed suitable to their gender specifically. Projective testing was one area which was recognized as being relatively accessible to women, as testing occupied a lower status (Bohan, 1990). However, once clinical Psychology and projective tests gained some element of prestige, for example following the World Wars and the International Rorschach Congress, the proportion of men in those areas increased, and women were less likely to be in positions on committees. Later, the legitimacy of projective tests became highly questioned and their prestige was soon to drop in both Britain and the US.

Rorschach criticism was apparent very near the beginning of its introduction to Britain (e.g., Lewis, 1934, see Hubbard & Hegarty, 2016) and those at the Maudsley Hospital were at the forefront of this criticism. In 1942 Lewis employed Hans Eysenck in the role of Senior

Research Officer. At the Maudsley, Eysenck oversaw the training of the first British clinical psychologists at the Institute of Psychiatry (Buchanan, 2010). In the early days, the Rorschach was taught by Swiss expert Maryse Israel to trainee psychologists, however, this training was discontinued as early as 1955. A 'critical discussion' meeting was held on Saturday May 21, 1955 and all members of the Committee of Professional Psychologists were invited to give their comments on this decision. The discussion appears to have done nothing but confirm their anti-Rorschach position. The Maudsley training program dominated clinical teaching in Britain producing twice as many graduates as the Tavistock, whose courses declined further in the 1970s (Buchanan, 2010). Eysenck's position at the Maudsley undoubtedly impacted the institution's approach to projective techniques. In 1959 he wrote a review and concluded that "the Rorschach has failed to establish its scientific or practical value" (Buros, 1959, p. 277).

As projective methods' popularity began to wane, other tests gained in popularity, especially those with a statistical underpinning. More 'objective' standardized tests (named according to the *Mental Measurements Yearbook* projective/objective dichotomy), such as cognitive and IQ tests increased in use as the use of projective tests decreased (Buchanan, 1997). From the 1950s there was a growth in concerns surrounding validity, reliability and the statistical nature of tests (e.g., Cronbach & Meehl, 1955). This was particularly exemplified in the development of the Diagnostic Statistical Manual from 1952 (DSM, Grob, 1991), and the American Psychological Association's 1954 attempt to standardize the Rorschach. In 1961 the MMPI overtook the Rorschach as the most popular psychological test in the US (Buchanan, 1997).

In Britain, the Standing Committee on Test Standards was established by the BPS in 1980 and investigated test popularity amongst psychologists as the use of psychological testing began to come under social scrutiny (Tyler & Miller, 1986). Findings showed that cognitive/intelligence tests were the most commonly used, followed by achievement/attainment tests, then personality tests, and finally developmental tests. For personality tests, questionnaires were the most popular measure, with attitude measures and personal construct measures following a close second. Projective tests came in third place. Overall, in the responses the Rorschach was criticized more often than it was supported.

Therefore, just as women had been viewed as not suitable for Psychology earlier in the century, the Rorschach and other projective techniques were similarly positioned as such at the end of the century. Beliefs about what can be considered 'legitimate,' or reliable and valid knowledge in Psychology has changed historically. In a broad sense, the actions of Psychology have delegitimized groups of people as well as the tools used by (some) psychologists in order to provide an impression of legitimacy and to imply a striving toward 'truth' and objective science. This is highly important considering the problematic nature of Psychology's past and present. Psychology has been deeply implicated in histories of eugenics, racism, colonialism, homophobic and transphobic practices and, as we have gone only a small way to show, sexism. What psychologists say about different groups of people and the tools psychologists use *does* a great deal—especially to those people being talked about (see Hubbard & Hare, 2015). In this next brief example, we wish to illustrate how sometimes problematic practices in Psychology's past have been disrupted by the actions of women wielding Rorschach cards, showing how these histories are not just similar, but have also clashed into one another at certain points in time.

## Ink Blots + Statistics = Evelyn Hooker and the 'Overt Male Homosexual'

In 1957 Evelyn Hooker published 'The Adjustment of the Overt Male Homosexual' in *The Journal of Projective Techniques.* It was also published in the *Mattachine Review*—the magazine of the Mattachine Society, the gay organization from which Hooker had recruited many of her gay participants. Hooker began this study in 1953 upon the request of her gay friend and previous student Sam From (see Minton, 2002 for a full account of Hooker's work and role in US emancipatory science). Hooker tested 60 men—30 gay and 30 straight—using the

Rorschach ink blot test. Men from each of the two groups were paired, being matched for age, intelligence and IQ. Each pair of Rorschach responses were then anonymized and Hooker asked two Rorschach clinicians, including the Rorschach expert Bruno Klopher with whom she worked with at UCLA, to report back which response was from the gay participant and which was from the straight participant. She found that despite the fact that the Rorschach was being used to diagnose 'homosexuality' (as this was a considered a mental illness by the APA and included in the DSM until 1973), these clinicians could not, above the level of chance, distinguish between the two groups. This work therefore seriously drew into question the legitimacy of 'homosexuality' being considered a clinical entity and psychological illness. Hooker's work both in the 1950s and later, including her role on the 'Task Force on Homosexuality,' was shown to be pivotal in the shift of attitudes in Psychology about the mental health of queer people (see Minton, 2002).

What is additionally interesting given the focus of this chapter, is how Hooker's 1957 paper depended on the legitimacy of the Rorschach. Without the Rorschach being considered a legitimate reliable psychological test at the time of the study, this paper could not have been so effective in motions to remove the pathologization of 'homosexuality.' Without belief that it was a reliable and valid test for detecting homosexuality, the results would have been meaningless.

Hooker utilized statistical methods to demonstrate that the clinicians' readings of Rorschach responses were no better than chance. Hegarty (2003b) specifically discussed Hooker's use of significance testing: Hooker's conclusion that there was no difference between the gay men and straight men's Rorschach responses might hinge on whether Hooker conducted independent or paired sample t-tests. Hooker argued that in a clinical setting psychologists would not receive two matched-paired Rorschach responses. She therefore did an unmatched analysis. Had she has chosen otherwise, it might have been possible for her to argue that there *were* distinguishable differences between the Rorschach responses of the gay men and straight men. Hegarty (2003b) highlighted how gay men gave more distinct responses than straight men, and also had more of the 'gay signs' according to Wheeler's (1949) signs for identifying gay men. Whether these can be considered 'significant' or not however, utterly depends on how we interpret statistics. Statistics require interpretation just as interpreting ink blots does; perhaps the most obvious example being the chosen significance level of 0.05. This point at which results are described as 'significant' or not, is subjective—it's a chosen point agreed upon by the discipline. To quote Hegarty (2003b):

> ...significance testing is an inexact process, and that the means by which marginally significant results are determined to be 'significant' or 'non-significant' forms part of the historical process by which scientific 'facts' about sexuality are constructed. (p. 31, see re-print 2018)

Hooker utilized the contemporaneous prestige of the Rorschach and statistics, from her suppressed positionality as a woman, to help gay men (see Hubbard, 2017; Minton, 2002). Despite the impact of this study such issues continue to permeate the discipline, though to a lesser extent. Historical accounts of women in Psychology tend to trace the same balancing act performed in Hooker's work; in order to claim a place in Psychology women have had to negotiate gender stereotypes about the fragility of women alongside the dominance of positivist epistemology.

Alternative models of prestige and practice could be drawn from existing feminist psychological scholarship, which utilizes a broader spectrum of methodologies and epistemologies. Qualitative work in particular has previously been devalued as being too subjective, and therefore its use within feminist Psychology becomes associated with gender stereotypes of emotionality and assumptions of subjectivity/political investments (Shields, 2007). This narrative is complicated by the value of the personal-political in feminist thought generally, and in feminist standpoint theory specifically (Harding, 1986). Feminist work is more often reflexive, qualitative and from experience, and less likely to have passive voice expectations in writing. Feminist psychologists are forced to navigate critiques of 'insider' research; the 'hybrid insider/outsider position' adopted by many feminist and minority researchers offers valuable insights, yet continues to be devalued by hegemonic psychological science (Wilkinson & Kitzinger, 2013). The same is true for scholarship conducted by other marginalized groups; decolonial scholarship documents similar processes by which the perspectives of Global Majority (non-Western) people are devalued (Mohanty, 2003).

## Conclusion

In highlighting the androcentric history often told about Psychology and indicating how Psychology's tools are also implicated in social beliefs about what is 'legitimate' we hope to show the value and importance of feminist history. Here, we are not attributing legitimacy to types of psychologists or particular methods (be they ink blots or statistics). We instead explore how historically the discipline has attributed characteristics such as 'subjective' and 'legitimate' to different genders and to different tools. In doing so, we demonstrate how beliefs about these things are mirrored in wider societal beliefs about gender and what 'objective science' should look like. In taking a particular feminist perspective, we note how women were at a particular disadvantage, as were other marginalized groups (and especially those women who embodied a variety of marginalized identities). Consideration of these intersections is vital when conducting historical and critical analysis of Psychology's past to avoid re-telling, or emulating, histories which have uncritically concentrated on the stories of white straight middle and upper-class cisgender men.

Throughout this chapter, we have attempted to give a short analysis of power using a few examples of gendered power in the history of Psychology. Dynamics of power and marginalization permeate who has been allowed to become psychologists and what tools are available to them. It is important to remember that as we discuss the powers at work within Psychology, that we must keep sight of why these dynamics matter—because Psychology's power flows outwards, as well as inwards. The processes of exclusion, stereotyping, and epistemic violence described in our chapter have effects far beyond the careers and wellbeing of women in Psychology. For this reason, it is imperative that we uncover and disrupt Psychology's power for our science to be both equitable and ethical.

## References

Amâncio, L., & Oliveira, J. M. (2006). IV. Men as individuals, women as a sexed category: Implications of symbolic asymmetry for feminist practice and feminist psychology. *Feminism & Psychology, 16*(1), 35–43.

American Psychological Association. (2000). *Women in academe: Two steps forward, one step back*. Washington, DC: APA.

Bernstein, M. D., & Russo, N. F. (1974). The history of psychology revisited: Or, up with our foremothers. *American Psychologist, 29*(2), 130.

Bohan, J. S. (1990). Contextual history. A framework of re-placing women in the history of psychology. *Psychology of Women Quarterly, 14* (2), 213–227.

Brunner, J. (2001). "Oh those crazy cards again": A history of the debate on the Nazi Rorschachs, 1946–2001. *Political Psychology, 22*(2), 233–261. https://doi.org/10.1111/0162-895X.00237.

Buchanan, R. D. (1997). Ink blots or profile plots: The Rorschach versus the MMPI as the right tool for a science-based profession. *Science, Technology and Human Values, 22*(2), 168–206.

Buchanan, R. D. (2010). *Playing with fire: The controversial career of Hans J. Eysenck*. New York: Oxford University Press.

Burfoot, A. (2016). *First ladies of running*. New York: Rodale.

Buros, O. K. E. (Ed.). (1959). *The fifth mental measurements yearbook*. Highland Park, NJ: Gryphon Press.

Causadias, J. M., Vitriol, J. A., & Atkin, A. L. (2018). Do we overemphasize the role of culture in the behavior of racial/ethnic minorities? Evidence of a cultural (mis) attribution bias in American psychology. *American Psychologist, 73,* 243–255.

Cronbach, L., & Meehl, P. (1955). Construct validity in psychological tests. *Psychological Bulletin, 52*(4), 281–302.

Fancher, R. E., & Rutherford, A. (2012). *Pioneers of psychology*. New York: W. W. Norton.

Furumoto, L. (2003). Beyond great men and great ideas: History of psychology in sociocultural context. In *Teaching gender and multicultural awareness: Resources for the psychology classroom* (pp. 113–124). Washington, DC: American Psychological Association.

Furumoto, L., & Scarborough, E. (1986). Placing woman in history of psychology: The first American women psychologists. *American Psychologist, 41,* 35–42.

Grob, G. N. (1991). Origins of DSM-I: A study in appearance and reality. *American Journal of Psychiatry, 148*(4), 421–431.
Gergen, K. J. (1973). Social psychology as history. *Journal of Personality and Social Psychology, 26*(2), 309–320.
Gasser, C. E., & Shaffer, K. S. (2014). Career development of women in academia: Traversing the leaky pipeline. *Professional Counselor, 4*(4), 332–352.
Guthrie, R. V. (1990). Mamie Phipps Clark (1917–1983). In A. N. O'Connell & N. F. Russo (Eds.), *Women in psychology: A bio-bibliographic sourcebook* (pp. 66–74). New York: Greenwood Press.
Hall, J. (2007). The emergence of clinical psychology in Britain from 1943 to 1958, Part II: Practice and research traditions. *History and Philosophy of Psychology, 9*(2), 1–33.
Haraway, D. (1988). Situated knowledges: The science question in feminism and the privilege of partial perspective. *Feminist Studies, 14*(3), 575–599.
Harding, S. (1986). *The science question in feminism*. London: Cornell University Press.
Harding, S. (1992). Rethinking standpoint epistemology: What is "strong objectivity?" *The Centennial Review, 36*(3), 437–470.
Hegarty, P. (2003a). Homosexual signs and heterosexual silences: Rorschach research on male homosexuality from 1921 to 1969. *Journal of the History of Sexuality, 12*(3), 400–423.
Hegarty, P. (2003b). Contingent differences: A historical note on Evelyn Hooker's uses of significance testing. *Lesbian and Gay Review, 4*(1), 3–7.
Hegarty, P. (2007). Getting dirty: Psychology's history of power. *History of Psychology, 10*(2), 75–91.
Hooker, E. (1957). The adjustment of the male overt homosexual. *Journal of Projective Techniques, 21*(1), 18–31.
Hubbard, K. (2017). Treading on delicate ground: Comparing the lesbian and gay affirmative Rorschach research of June Hopkins and Evelyn Hooker. *Psychology of Women Section Review, 19*(1), 3–9.
Hubbard, K. (2018). Queer signs: The women of the British projective test movement. *Journal of the History of the Behavioural Sciences, 53*(2), 265–285.
Hubbard, K., & Hare, D. (2015). Psychologists as testers. In J. Hall, D. Pilgrim, & G. Turpin (Eds.), *Clinical psychology in Britain: Historical perspectives*. History of Psychology Centre Monograph No. 2. British Psychology Society. Leicester: Blackwell.

Hubbard, K., & Hegarty, P. (2016). Blots and all: A history of the Rorschach ink blot test in Britain. *Journal of the History of the Behavioral Sciences, 52*(2), 146–166.

Hyde, J. (1990). Meta-analysis and the psychology of gender differences. *Signs, 16*(1), 55–73.

Lemov, R. (2011). X-rays of inner worlds: The mid-twentieth-century American projective test movement. *Journal of the History of the Behavioral Sciences, 47*(3), 251–278.

Lewis, A. J. (1934). Melancholia: A clinical survey of depressive states. *The British Journal of Psychiatry, 80,* 277–378.

Madera, J. M., Hebl, M. R., & Martin, R. C. (2009). Gender and letters of recommendation for academia: Agentic and communal differences. *Journal of Applied Psychology, 94*(6), 1591–1599.

McCarthy Woods, J. (2008). The history of the Rorschach in the United Kingdom. *Rorschachiana: Journal of the International Society for the Rorschach, 29*(1), 64–80.

Mignolo, W. D. (2011). *The darker side of western modernity: Global futures, decolonial options.* London: Duke University Press.

Milar, K. S. (2000). The first generation of women psychologists and the psychology of women. *American Psychologist, 55*(6), 616–619.

Minton, H. L. (2002). *Departing from deviance: A history of homosexual rights and emancipatory science in America*. London: University of Chicago Press.

Mohanty, C. T. (2003). *Feminism without borders: Decolonizing theory, practices solidarity*. Durham: Duke University Press.

Morawski, J. G., & Agronick, G. (1991). A restive legacy: The history of feminist work in experimental and cognitive psychology. *Psychology of Women Quarterly, 15*(4), 567–579.

Rees, T. (2011). The gendered construction of scientific excellence. *Interdisciplinary Science Reviews, 36*(2), 133–145.

Richards, G. (2002). *Putting psychology in its place: A critical historical overview.* Hove, East Sussex: Psychology Press.

Rorschach, H. (1921). *Psychodiagnostics: A diagnostic test based on perception* (P. Lemkau & B. Kronenberg, Trans.). New York: Grune & Stratton.

Rovenpor, D. R., & Gonzales, J. E. (2015, January). Replicability in psychological science: Challenges, opportunities, and how to stay up-to-date. *Psychological Science Agenda, 29*(1). Retrieved from https://www.apa.org/science/about/psa/2015/01/replicability.aspx.

Rutherford, A., Vaughn-Blount, K., & Ball, L. C. (2010). Responsible opposition, disruptive voices: Science, social change, and the history of feminist psychology. *Psychology of Women Quarterly, 34*(4), 460–473.
Rutherford, A., Vaughn-Johnson, E., & Rodkey, E. (2015). Does Psychology have a gender? *The Psychologist, 28*(6), 508–510.
Scarborough, E., & Furumoto, L. (1989). *Untold lives: The first generation of American women psychologists*. New York: Columbia University Press.
Shields, S. A. (2007). Passionate men, emotional women: Psychology constructs gender difference in the late 19th century. *History of Psychology, 10*(2), 92–110.
Spears, R., & Smith, H. J. (2001). Experiments as politics. *Political Psychology, 22*(2), 309–330.
Stevens, G., & Gardner, S. (1982). *The women of psychology*. Cambridge, USA: Schenkman Publishing.
Tyler, B., & Miller, K. (1986). The use of tests by psychologists: Report on a survey of BPS members. *Bulletin of the British Psychological Society, 39*, 405–410.
Valentine, E. R. (2008). To care or to understand? Women members of the British Psychological Society 1901–1918. *History and Philosophy of Psychology, 10*(1), 54–65.
Valentine, E. R. (2010). Women in early 20th-century experimental psychology. *The Psychologist, 23*(12), 972–974.
Wheeler, W. M. (1949). An analysis of Rorschach indices of male homosexuality. *Rorschach Research Exchange and Journal of Projective Techniques, 13*(2), 97–126.
Wilkinson, S., & Kitzinger, C. (2013). Representing our own experience: Issues in "insider" research. *Psychology of Women Quarterly, 37*(2), 251–255.
Wylie, A. (2004) Why standpoint matters. In S. G. Harding (Ed.), *The feminist standpoint theory reader: Intellectual and political controversies* (pp. 339–352). Hove: Psychology Press.

# 14

# Psychology in Times of Smart Systems—Beyond Cyborgs and Intra-action

Ines Langemeyer

## Puns, Metaphors, and Allusions

In postmodern theorizing, the cyborg metaphor stands for the displacement of boundaries between humans and the 'nonorganic' and 'artificial' materiality. By advocating a symmetry of these matters, the cyborg is something more than a symbol. It is an attempt to taper the epistemic problem that also "the boundary between science fiction and social reality [would be] an optical illusion" (Haraway, 1985, p. 191). For Donna Haraway, the feminist luminary of postmodernism, the metaphor of the cyborg thus condenses (or "diffracts" as she later puts it)[1] the instability

[1]"Diffraction does not produce 'the same' displaced, as reflection and refraction do. Diffraction is a mapping of interference, not of replication, reflection, or reproduction. A diffraction pattern does not map where differences appear, but rather maps where the *effects* of differences appear" (Haraway, 1992, p. 300).

I. Langemeyer (✉)
Institut für Allgemeine Pädagogik, Karlsruhe, Germany
e-mail: ines.langemeyer@kit.edu



K. C. O'Doherty et al. (eds.), *Psychological Studies of Science and Technology*, Palgrave Studies in the Theory and History of Psychology,
https://doi.org/10.1007/978-3-030-25308-0_14

in all current forms of life as "a struggle over life and death" (ibid.). Allusions to the 'to-be-or-not-to-be' question indicate the scope of intervention.

In more rational words, Haraway wants scientists to recognize themselves not only as observers, as neutral witnesses, so to speak, but simultaneously as actors related to the world and responsible for it. Her way of speaking literally is not just a cryptic pun or a wink (e.g.: "Unlike the hopes of Frankenstein's monster, the cyborg does not expect its father to save it", Haraway, 1985, p. 192). It is deliberately tongue-in-cheek, playing with the subject matter. Allusions to the Hamlet-drama, like the ambiguous comment that "there was always the specter of the ghost in the machine" (Haraway, 1985, p. 193), indicate that Haraway tries to create "a condensed image of both imagination and material reality", and to find "pleasure in the confusion of boundaries and for responsibility in their construction" (Haraway, 1985, p. 191). She uses this pleasure to tackle the 'normal', 'clean' ways of reporting, perceiving, and investigating empirically. To reflect an object of study theoretically should thus become a question of responsibility and self-entanglement.

While recognizing and keeping up this aim, this kind of critical thinking shall be put to a test. By simultaneously clarifying its theorems and methods, the underlying methodology is applied to today's contradictions in the field of *learning* into which more and more technologies are inscribed. The feminist techno-scientific critique cannot be reduced to the work of Haraway, therefore more recent continuations by Karen Barad are also considered.

## Tackling the Realist Epistemic Belief

In her continuations of this feminist approach, Karen Barad elaborates further on Haraway's idea to see boundaries as ultimately "lived relations of domination" (Haraway 1985, p. 194). To do so, Barad (2007, p. 42) accentuates "semiotic and deconstructivist positions". She also takes sides with praxeology, in that "knowing does not come from standing at a distance and representing but rather from *a direct engagement with the world*" (p. 49). Like Haraway's cyborg metaphor,

Barad thus emphasizes the transformative forces that come from *within* socio-technological relations rather than from outside. She coins this 'within' "intra-action". She draws attention to the cyborg-like penetration of humans and artificial entities where matters "intra-act" and ontological distinctions become iridescent. Her argument regarding this "intra-action" is developed as follows.

Barad takes the historical development of scientific disciplines, including laboratory research, as given. She argues that laboratory research works with "a rigid apparatus with fixed parts" and thus creates a certain meaning for the "notion of 'position'" (Barad, 2003, p. 814). For this reason, she concludes that "any measurement of 'position' using this apparatus cannot be attributed to some abstract, independently existing 'object', but rather is a property of the *phenomenon*" as a whole. This implies changes in the notion of a phenomenon. Barad assumes a "causal relationship between the apparatus of bodily production and the phenomena produced" (Barad, 2003, p. 814). This relationship would be "one of 'agential intra-action'" (Barad, 2003, p. 814). Intra-action expands the notion of a phenomenon to be inherent to the apparatus (as a technologically controlled and designed practice), because it is both the carrier of causality (experimental reactions) and of the particular appearance (experimental results). There is no 'outside' where the observer could stand and thus not a neutral-witness-position to occupy. Concomitantly, Barad makes a certain epistemic belief a subject of discussion: The realist belief which implies, among others, that the epistemic subject exists independently from the object and, vice versa, that the object 'out there' is recognizable by the subject, because the disturbing insight from quantum physics, the Heisenberg uncertainty principle,[2] displays "material exclusion of 'position' and 'momentum' arrangements" or, in more common language, questions the

[2]Heisenberg reflected on the inexactness of measurements in physics. The disturbance of an electron by a photon would be necessary for measuring the electron's movement, but at the same time the photon adds energy to the system the electron is part of. Therefore, an observation without changing the object to be observed would be impossible. This is what is meant by 'the Heisenberg uncertainty principle'. Simultaneously, it reflects the impossibility of measuring precisely both the position of the electron and its movement.

"epistemological inseparability of 'observer' and 'observed'": in Barad's own words, it problematizes *"the ontological inseparability of agentially intra-acting 'components'"* (Barad, 2003, p. 815).

Barad is not the first to recognize epistemic beliefs as an unconscious influence in the development of science. A predecessor of this thesis is Gaston Bachelard who investigated the realist belief as one of numerous "epistemic obstacles" (Bachelard, 2006). Barad (2011, p. 451) does not deal with epistemic obstacles but is rather interested in "why and how matters of science [...] are always already intra-actively entangled with questions of politics and power". This would ultimately lead to the core question: "Who and what gets excluded matters" (Barad, 2011, p. 451).

## Thus, Several Questions Emerge

1. In what ways are *intra*-actions different from *inter*-actions and how do they bring new insights to the fore?
2. Is it justified to reinterpret the *epistemological* inseparability of observer and observed as an *ontological* rather than an apparatus-related inseparability?
3. In what ways is playing with an assumed symmetry of matter, especially of technological and human matter reasonable, and how does it lead to a radical critique with regard to societal life and action?
4. In what ways can we articulate issues of responsibility and exclusion in a more adequate and more substantial way?
5. How cogent is the feminist techno-scientific critique to solve the epistemological issues addressed?

With these questions, I put Barad's attempt to develop a new epistemic foundation at a distance. I am doubtful about the assumed immanence of societal problems of exclusion and domination within the epistemological problems raised in quantum physics. At first sight, Barad's argument seems to invoke a radical critique of empirical research. However, this argument stirs uneasiness about the usefulness of seeing the field of quantum physics (similar to Haraway's cyborg) not merely as a metaphor or an analogy for a thought-provoking epistemic problematic, but as immediately political. In addition, Barad's critique of the immanent epistemic position is presented as a general issue concerning *all* scientific research, whilst her argumentation refers to the realist belief of empiricism and positivism (which sees the representations and the entities to be represented as distinct and independent, Barad, 2003, p. 804). Thus, she ignores other paradigms, for example the dialectical one beginning with Marx's theses on Feuerbach. This is strange because both Barad and Haraway stand on the shoulders of Marxian thought. In what follows, I scrutinize whether Barad's entire argument is ultimately a short circuit, because it neglects the societal mediations that maybe justify why the findings of quantum physics reveal something about power relations in material societal life (cf. Langemeyer, 2017b).

To explain my doubts further: Like Haraway, Barad aims at understanding the workings of power. I agree with both, and especially with Barad, that we must pay attention to the unity of knowing and intervening (*Einheit von Erkennen und Verändern*). This becomes clear when she highlights that "the nature of power" lies in "the fullness of its materiality" (Barad, 2003, p. 810), in particular, in the "apparatuses [which, I.L.] are dynamic (re)configurings of the world, specific agential practices/intra-actions/performances through which specific exclusionary boundaries are enacted" (Barad, 2003, p. 816). To strengthen this argument, Barad reinterprets processes of signification by which phenomena become noumena: According to her, "meaning" would not be "ideational but rather specific material (re)configurings of the world" (Barad, 2003, pp. 818–819). However, this means that language as a practice recedes into the background and becomes a subordinate dimension of allegedly nonsymbolic material or, more specifically, technological (re) configurations of the world. This seems to exaggerate the theoretical intervention compared to Marx's plea to see the object of research as subjective practice to which significations belong. His first thesis on Feuerbach (Marx, 1947) accentuates subjectivity as concrete-sensual activity, knowing that subjective practice is not devoid of cognitive activity (not pure manual labor, so to speak) nor is cognition a simple effect of material practice. Otherwise, one should remember, puns, metaphors and allusions would be impossible.

Concerning the inseparability of observer and observed on the one hand, and the question of power as organized through an apparatus on the other, Barad draws on two arguments that somehow merge into one. The apparatus with its "intra-actions" and its epistemic power relations becomes the superior subject. Thus, she construes an immanence relation.

My way of questioning Barad's concept of materiality is motivated against the background of a critical appreciation of arguments presented by Louis Althusser, and later Michel Foucault, in structuralist French philosophy. In the early 1960s, when Althusser was interested in Bertolt Brecht's theater, he rejected the traditional notion of consciousness as something purely ideational and began to outline an analysis of power along the material practices of the 'apparatus' (cf. French also: *dispositif*), a concept he borrowed from Gaston Bachelard (Althusser, 1962, 2014). In more recent publications, Barad continues to work on the same issues as Althusser, such as the difference between homogeneous and historical time (Althusser, 2006). It is therefore astonishing not to find any reference to his work in this context (cf. Barad, 2017), while reference to Foucault's argument for structural immanence is made (Barad, 2007, p. 229; cf. pp. 199–204). Both Althusserian and Foucaultian analyses of power relations raised awareness of the lack of neutrality or impartiality of the scientific observer and the consequent missing distance for reflection. While this insight can be read as a general request for more critical reflection on power relations in science, in order to gain or regain relatively more distance from the apparatus, with Barad, this possibility recedes into the background. She neglects to discuss the ways in which societal subjects can enhance and expand the necessary conditions for themselves in order to regain a form of distance as practical and cognitive independence (including political independence)—shortfalls that she probably inherits from Althusser and Foucault.

Against these shortfalls, I draw on the work of Vygotsky, who is clearer about the necessity to develop scientificated societal relations as cognitive and practical empowerments.

## Rediscovering Vygotsky

Through Vygotsky's psychological approach, it is possible to discover that the problems addressed in the previous sections are not entirely new and have already been subject to considerable theoretical developments (for similar theoretical developments with Bertolt Brecht and Kurt Lewin, cf. Langemeyer, 2017b). Beyond realism, beyond inert ontological entities and self-reliant subjects and objects, Vygotsky has demonstrated the productivity of dialectical theorizing. The main features can be outlined as follows:

> Methodologically, a phenomenon should be studied in its most developed form by reconstructing how it emerges through previous forms (historico-genetic perspective), which is why the phenomenon should also be investigated in the process of its change (perspective on dynamics, mediations and transformations); the phenomenon of a developed form is then conceived of as a whole instead of isolated parts or elements, and the method applied needs to preserve the inner relations between the parts of a whole (holistic perspective); thus, the complexity of the objects of investigation is not reduced and the representations of these objects (theorems, concepts, or models) do not tend to feed false abstractions (structuralist, integral, or organic perspective); but since no method provides a guarantee for truth, it is necessary to reflect the process of theorizing and to determine the (historical) limits of scientific concepts, insights and generalizations (self-critical perspective) (cf. Langemeyer & Roth, 2006, p. 27).

This description is not meant to fixate the essence of dialectics; quite the contrary, it should help to recognize that dialectical thinking works by dissolving reified or reifying modes of thinking and transforming them into intellectual engagements with a changing world. These engagements are envisioned with emancipatory practice.

To explain the advantage of Vygotskian thought more concretely in relation to psychology: Vygotsky assumed that the human ontogenesis encompasses two lines of development which constantly interact so that neither one can be investigated immediately and isolated from the other. One would refer to the biological aspects of development, the other to the sociocultural aspects. Causal relationships could therefore not be referred to unidirectional impacts from one line onto the other. Thus, the permanent interaction of two lines of development means that the child growing up does not undergo a metamorphosis from a natural or biological to a societal being: it is always already both. Against this backdrop, a concrete observation of 'biological' and 'societal development' was rejected by Vygotsky. It therefore became possible only to refer to 'nature' or 'society' in an analytical and historical way. Similarly to the sociobiological development, Vygotsky also assumed interactions between the individual and the collective level in human development, which is why he clearly rejected methodological individualism (where everything emerges from the individual level and must be studied from the allegedly simple individual forms to the complex societal forms), and structural determinism (where individuals are merely an effect of societal structures). In contradistinction, Vygotsky defined as a law of psychic development, that "every function in the child's cultural development appears twice: first, on the social level, and later, on the individual level" (Vygotsky, 1978, p. 57). This contributed to his methodological insight that studying the genesis of higher psychic functions can be accomplished properly only by investigating them "first as a collective form of behavior, as an inter-psychological function" and then "as an intra-psychological function, as a certain way of behaving" (Vygotsky, 1997, p. 95). Furthermore, the analysis by Vygotsky proceeds with reconstructions as to how the interiorized forms of action, as psychic means, contribute to forming more advanced psychic functions. The child's change in experiencing (and behavior) when he/she learns to think with scientific concepts rather than spontaneous or everyday concepts is a good example (Vygotsky, 1987).

Vygotsky therefore saw a problem in epistemic beliefs, virulent now as then, that empirical investigations of isolated phenomena would suffice, and that concepts, without peril and ambiguity, would serve to capture their truth. Against these beliefs, his understanding was that effective psychological research would require theoretical reflection and therefore a methodology of an "indirect" way, which implies that observation in the traditional sense of the realist position is impossible, and that the mediation that theories organize should be methodologically controlled. It is for this reason that, drawing on Marx, he argued:

> After all, if concepts, as tools, were set aside for particular facts of experience in advance, all science would be superfluous; then a thousand administrator-registrators or statistician-counters could note down the universe on cards, graphs, columns. Scientific knowledge differs from the registration of a fact in that it selects the concept needed, i.e., it analyzes both fact and concept. (Vygotsky, 1997, p. 251)

Vygotsky's approach thus became recognized as a new and original epistemology of psychology (cf. Friedrich, 2012). Like Haraway later, he worked with the insight that the researcher's subjectivity is always already a product of human activity and its societal contexts. This similarity is not a surprise, as already mentioned: both Haraway and Vygotsky were building on Marx's first thesis on Feuerbach. This clarifies their commonalities with regard to the epistemic argument that the researcher's subjectivity does not 'witness modestly' objectivity (cf. Haraway & Goodeve, 2018), but that it is always already entangled with the same matter as psychological research: the origins and the modes of conscious behavior. But differently from immanence philosophy, Vygotsky saw scientific consciousness raised only along with critical work on concepts, i.e., with the struggle for cognitive independence to deliberately investigate things anew from another perspective, and not to reify the products of a critical engagement with practice (Vygotsky, 1997, p. 251).

Cultural-historical development as a subject matter (how people produce their lives, how they find meaning in it etc.) is—similarly to psychological development—not seen as something immediately observable and requires dialectical theorizing. It is mainly for this insight that Vygotsky must be seen as a Marxian scholar (cf. Ratner & Silva, 2017; Sève, 2018; Stetsenko, 2016, p. 183). The history of scientific concepts and ways of doing science do not exist independently from other practices in societal life, which implies that they do not exist independently from biological and other material processes either (Schraube, 2009). To avoid false ontological divisions, separations and dissections which distort the subject matter into unrecognizable parts, Vygotsky elaborated on holistic methodological considerations to find adequate "units of analysis" which "[make] it possible to see the relationship between the individual's needs or inclinations and his thinking" or "the relationship that links his thoughts to the dynamics of behavior, to the concrete activity of the personality" (Vygotsky, 1987, pp. 50–51).

Against the naïve expectation of a positivistic realism, Vygotsky's approach conveyed that truth would not be immediately available through techniques (or scientific methods) of observation and so accessible merely by fixations of the research object. He forged the wisdom for a dialectical methodology that "it is only in movement that a body shows what it is" (Vygotsky, 1978, p. 65). But unlike Barad, Vygotsky suggested taking the dialectical interaction between the special sciences and the general as essential:

> [...] we can give no absolute definition of the concept of a general science [...,] it can only be defined relative to the special science. From the latter it is distinguished not by its object, nor by the method, goal, or result of the investigation. But for a number of special sciences which study related realms of reality from a single viewpoint it accomplishes the same work and by the same method and with the same goal as each of these sciences accomplish for their own material. (Vygotsky, 1997, p. 249)

The perspective of the "special science" thus captures what feminist philosophers like Haraway find in 'situated' knowledge and subjectivity: It includes a concrete researcher subject and a concrete research object, located in concrete practice and in a particular arrangement of scientific investigation. Its result is concrete (not abstract) experience, fueled (more or less) by the materiality of societal contradictions and conflicts. But since reflection uses generalized meaning, and generalizations can be trapped in illusions or blurred imaginations, general science comes into play. "General science" is any self-critical philosophical engagement with the particular discipline in which researchers develop their thoughts and insights. As Vygotsky clarifies, this way of doing science as general science is still dependent on the same objects, methods, and experiences. However, it does not assume that researchers' consciousness remains an immanent effect of material relations:

> We have seen that no science confines itself to the simple accumulation of material, but rather that it subjects this material to diverse and prolonged processing, that it groups and generalizes the material, creates a theory and hypotheses which help to get a wider perspective on reality than the one which follows from the various uncoordinated facts. [...] When the material is carried to the highest degree of generalization possible in that science, further generalization is possible only beyond the boundaries of the given science and by comparing it with the material of a number of adjacent sciences. This is what the general science does. (Vygotsky, 1997, p. 249)

The relative distance from research objects that scientific research needs is thus neither given nor can it be assumed to be stable like an ontological fact. Distance from the object of study is achieved by moving between special and general sciences and by reflecting the different experiences each science enables. Turning to general science is not an end in itself, but is necessary when false abstractions take the lead.

Returning to the five questions, some answers may be given now:

The assumption that intra-actions bring more or better insights than interactions is only striking when we know the apparatus which frames intra-actions. We need to make the apparatus (like the particular laboratory research design or, more generally, the political regime in which we live) an object of study in order to understand how it produces intra-actions from within. Researchers must expand their questioning and scrutiny from the original concrete object of study to concrete research practice, its materiality, its representations, and its power relations. With Vygotsky, they need to shift the unit of analysis from the single positivistic objects of study to the apparatus of experiments, observations etc. I assume that Barad and Haraway agree largely with this conclusion.

However, I object to Barad's program that without deeper knowledge of the apparatus, intra-actions are not differently intelligible than interactions. That means that the problem of locating the object of research at a distance to observe it is only postponed: Research would have to start with the apparatus. But then the problem is to grasp the complex apparatus. Its scientific observation is not easier than the investigation of the phenomenon it produces. The alternative I suggest as a Vygotskian scholar is to strengthen the self-critical relation of the researcher toward her/his own concepts and the generalizations used. To give an example:

Barad construes responsibility in relation to concrete materiality when she speaks about "material reconfigurations of spacetimemattering" (Barad, 2017, p. 63). These reconfigurations are seen as caused by radioactivity after nuclear bombs had destroyed Hiroshima and Nagasaki, for instance, and had contaminated the soil. To underline its agential role as a causer of diseases, radioactivity appears grammatically as a subject-like human entity while the human societal actors fall out of sight. This interpretation becomes excessive when radioactivity also seems to do the epistemic work of reworking notions and calculations (Barad, 2017, p. 63). In line with this rhetoric, Barad considers objects of study in physics as responsible for colonialist worldviews and endeavors:

> The void occupied a central place in Newton's natural philosophy. [...] The void, in classical physics, is *that which literally doesn't matter*. It is merely that which frames what is absolute. While the so-called voyages of discovery, bringing data (including astronomical and tidal changes) culled from European journeys to non-European sites aided Newton in his efforts to develop a natural philosophy that united heaven and earth, Newtonian physics helped consolidate and give scientific credence to colonialist endeavors to make claims on lands that were said to be de-void of persons in possession of culture and reason. (Barad, 2017, p. 77)

While I do not question the primary concern of Barad's critique (the need for awareness of injustice and responsibility), I see her argumentation here lacking a dialectical turn—it is stuck in a rather unhistorical reflection of materiality: Are the objects of physics (like radioactivity, matter, the void etc.) really imposing their ways of 'mattering' onto the understanding of philosophers so that their interpretations could serve the imperialist regimes of Europe as a legitimization for compulsory acquisition of land, or: Isn't it more convincing to say that Newton's philosophy and physics is implicitly influenced by colonialism, his own time and context, when he theorized the relation between matter and void?

With regard to the remaining three questions concerning the layout of a radical critique, detailed answers are not yet clear. As mentioned above, I am doubtful that the assumed equivalence or symmetry of human subjects and nonhuman objects would revolutionize the point of departure for a radical critique, hence the question of responsibility tends to be leveled out. Symmetry is not an adequate notion of the human–world relation to clarify that human beings depend in their development and well-being on the care of others, on participating in societal practices, foresight, security and freedom, whilst the world 'out there', especially the diversity of species, could probably exist more easily if they were unaffected by humankind. It is not physics, but *societal* practices (including sciences) that reconfigure matter and thus the conditions of individual lives, which means that a number of mediating instances need to be taken into account. Although playing with the assumed symmetry is supposed to reveal, among others, problems of colonialism and injustice as they occur in situ, the effect on critical thought might be altogether rather contrary to this. Therefore, these problems shall be subject to discussion in the next sections.

## Human or Artificial Intelligence?

Similarly to Vygotsky, Barad's feminist techno-scientific approach tries to open up "deeper understanding of the ontological dimensions of scientific practice" (Barad, 2007, p. 42). In what follows, I put these approaches to a test: In what ways do they improve practices of doing science when one starts questioning whether, for example, human "intelligence" is essentially human, or no longer exclusive to human brains? For Haraway and Barad, the crucial point of supporting their view of cyborgs as well as intra-actions of an apparatus as the central matters of theorizing, or not, is whether this contributes to a better articulation of the question of responsibility in social practice and of coming to a better understanding of what that means in terms of social justice.

Undeniably, the ontological misunderstanding of the exclusiveness of human skills is boosted nowadays not only by puns: When companies sell their products as 'intelligent' systems, as 'smart' tools, and the like, they create meaning that blurs the ontological distinction between technology and human intelligence. From a psychological point of view, the fundamental question is however, whether concepts like cognition, awareness, perception, action, motivation, and judgement are therefore still unproblematic or should be rethought, since cognitive capacities are no longer considered as belonging solely to the individual psyche but exceeds it through technological devices or organization systems.

Unlike Haraway and Barad, I suggest interpreting these changes not as an ontological shift, but as a historically new quality of societal life. This implies seeing the challenge for psychology not merely in the transgression of boundaries between human bodies and 'nonorganic' and 'artificial' matter (I would include this as part of the entirety of cultural human development), but rather in understanding the particular cultural (or collective) development of the psychic functions in the light of a technologically driven *societal* process of the *scientification of capacities to act*.

Scientification is understood as a rather precarious process of non-simultaneous and nonlinear cultural human development. One aspect of this development is the world-changing character of scientific inventions such as computers and the internet. More and more dimensions of societal life are dependent on these scientifically invented technologies. The everyday culture (the communication with others, reading, and writing, self-reflection) has changed tremendously. Furthermore, changes in labor and learning concretize the constraints and challenges that individuals face in our lives. Regarding demands of qualification, the scientification process is not necessarily clear or unambiguous: There is no automatism that means the individual worker becomes a scientist just because technologies are produced scientifically. However, given digitalization, the main aspects of the development of labor lie in the intellectualization and scientification of work: i.e., relevant intervention into digitalized processes is only possible via the apt use of computers and scientific methods (Langemeyer, 2017a).

To develop a deeper understanding of this matter of "smart technology", it is firstly shown that the perspective of "agential realism" which Barad proposes lacks some relevant presuppositions. The presuppositions become relevant with regard to new forms and constellations of technological power, such as autopilots and completely automated systems that make more and more decisions concerning peoples' lives. With regard to this automation, responsibility comes into play through intellectualized working capacities. These require not only the acquisition of scientific stores of knowledge but the competences to imagine and anticipate the problems and risks that can be triggered or unleashed within the societal use of IT systems and their connections to other systems and contexts.

'Smart' houses, offices, production plants, clinics, and even entire cities are in the making, and some are already tested in reality. In this context, the technologies of 'deep learning' and 'organic computing' have obtained the capacity to transform themselves independently while processing data or while they interact with the environment. These technologies are seen, for instance, as powerful inventions in accomplishing the transition from nuclear and fossil powers to the so-called sustainable energies. Another domain is the calculation of risks. Politics concerned with climate change or biopolitics, insurance companies, financial institutions, stock exchange, personnel recruitment departments, etc., have become interested in modeling, simulating and forecasting by means of computational analytics. If their results are fed into further automatic processes, the contribution of human intellectuality to this no longer seems important or necessary. However, an increasingly self-referential technological apparatus means that political will for interventions or shifts cannot be formed and critically developed in a timely manner. If software takes care of processing people's annual tax declarations, and if this software 'learns' through processing the data fed into its system, then certain exceptions to rules will not be found against the background of considerations of societal responsibilities and ethics, but only in relation to the data and the patterns extracted from it. Reasoning related to *responsibility* would still require the involvement of human reflection, i.e., to evaluate people's needs and futures in the light of the actual decisions. However, human involvement in processing data is increasingly systemically excluded, and an intervention into conclusions automatically 'drawn' by the software would firstly demand a reconstruction of the applied calculations. This can imply a delay of insight and decisions with severe consequences. This situation likely reminds us of having a ghost in the machine (cf. Knorr Cetina, 2007).

The semantic dimension of this societal change can be interpreted as follows. On the one hand, the economic and political vision behind this digitalization is, as it were, traditional or old-fashioned: Human subjectivity is seen as outdated and not fail-safe, whilst advanced technologies bring human capacities, physically and intellectually, to perfection. On the other, what is experienced as a break is that not only manual skills and physical powers, but also perceiving, calculating, reasoning and—last but not least—learning, are considered as imperfect activities which are amendable through the most advanced technologies. It is ultimately the entirety of human subjective behavior that is reinterpreted as accessible by technological efforts and strategies, not only to replace it, but also to track and granulate it into individual-related data (Kucklick, 2014) in order to optimize it or, at least, to influence it. Objectives and criteria are set by companies, which acquire expertise from psychology and arts to sell their commercialized way of life. In contradistinction, the striving to develop personality and autonomy in their institutionalized forms as human rights therefore seems to be (or is deliberately construed as) a remnant of an outdated romantic vision of life.

## Is the Subject Matter of Psychology Transgressing to Technological Devices or Systems?

These developments of 'smart' technologies and cyber systems are not trivial to psychological theorizing, as the following issues may discern: Approaches of general psychology often presuppose that the acting subject (let's assume: sober-minded and without defects) is responsible for the consequences of her/his actions. At least, they accept the societal conditions which ascribe responsibility to human actors. The conditions under which their actions take place are usually acknowledged theoretically as possibly 'restraining', 'intervening', or 'convening'. The "attribution theory" for instance, distinguishes between "internal" and "external" reasons for either success or failure in actions that the individual attaches to him-/herself (Heider, 1958). Yet, what are "actions" conducted in a digital environment of cyber systems? And what about the "soft" power of algorithms and their "intra-actions", which could be seen in their self-referential transformations according to the data fed into them. If these material transformations are completely run by digital information, hard- and software, or if the system's interface provides users with suggestions, forecasts, judgements, and even with decisions, is the subject then still the center, or at least an instance of, responsibility? This is, on the one hand, a juridical issue (Are programmers or the owner to blame if an automated-driven car hurts someone?) and, on the other, elementary to psychological concepts such as "self-efficacy" or "self-coherence" and "personality development". Psychological approaches convey that experience with actions and personal responsibility contribute to forming a "self" with values and norms and certain ambitions in life. They also assume that well-being depends on experiencing oneself as the 'cause' or at least as the main driver of the respective actions (cf. Brandtstätter & Otto, 2009; Kuhl & Kaschel, 2004; Urhahne, 2008). In philosophy, George Canguilhem (2002, p. 68) relates the notion of health to someone's way of experiencing responsibility:

> I am well to the extent that I feel able to take responsibility for my actions, to bring things into existence and to create between them relations which would not come without me.

Similarly, Rahel Jaeggi argues for the experience that people overcome a "relation of relationlessness", thereby interpreting the opposite state with the concept of "alienation" (*Entfremdung*) in an anti-essentialist manner (2005). Yet, regarding the concrete technological uses of algorithms and cyber systems, the ratio between being the responsible agent or rather the string puppet is blurred. This problem is also cogent for learning.

Learning is often psychologically defined as a change in both behavior and cognition. But if algorithms start to make decisions, e.g., about someone's trajectory or progress with learning challenges: Is it then still the same cognitive and motivational process if learners stop bothering about interesting learning content ('What stirs my passions?') and aims ('What do others want me to do, and what is it that I want to do?'), about lessons learnt, goal-adequate issues, materials, and adequate forms of learning? Some might argue that learners who are using learning analytics are not necessarily prevented from taking responsibility for their learning progress. However, the inclination to avoid responsibility and effort while looking for a benefit from these technologies is already visible: People who use search engines usually have no insight into how their search request is processed. They might know that search engines automatically create user profiles to personalize the order of results. Yet, without being fully aware of the automated selection done by the search engine and its logics of prioritizing, the individual user consumes the information given—and with it, accept the priorities set. This is not trivial, as people implicitly infer from such synesthetic impressions what is of higher importance, or rather what is irrelevant. If social media makes suggestions about 'friends' or 'colleagues' to keep contact with, these decisions receive their clues not only from experiencing oneself in a certain situation but quite distinctly from algorithms.

This problem does not consist only in consumers' unreflected practice, but in the opacity of this technology itself: Not even programmers have a clear and complete understanding of the self-transforming algorithms they invented. In addition, if some information is given, the problem is not only that false or unreliable information needs 'good' information to reveal the error, the illusion or the lie: Often, digital information is not critically compared with *experience* as it can be *made* by humans. Thus the entire background of experiencing is alienated. The more that algorithms produce a societal reality that becomes the reference for interpretations and interpretative horizons, the more they undermine our self-critical engagements with concrete subjective practice. The distinction between objectified data and the subjective activity of experiencing is nowadays confused.

Digitalization implies a deep transformation of spaces in everyday life that seem to 'speak' immediately and innocently to us. Most clearly, this tendency can be experienced with technological devices equipped with bots or avatars, voice control, and audio interfaces so that one can interact with them in a pseudo-social form. Digital technology is designed to make sense of our activities and our life in general. This 'sense' is *made*, not merely in an interpretative manner as meaning, but also in a practical one as a mass of people are drawn into this 'machine' and (have to) supply more and more personal information to social media, for example, in order to receive benefits such as attention by 'friends' or followers, or storage space. Digital life has become a touchable and calculable form of life, often more attractive than its analogous correspondent. It reaches out to become *the main* way of being social. As the backbone of a new mode of automated production, digital technologies not only control other technologies but nowadays satisfy a number of needs for participation, belonging, and recognition.

By extending the possibilities of consuming 'sense' and by creating spaces of 'meaningful' activity, these technologies simultaneously minimize and distort the subjective activity of experiencing her-/himself. In the 1950s, when broadcasting and TV invaded the private sphere, Günther Anders criticized the new possibility of consuming pictures of other regions as well as information about situations and events elsewhere as a loss of the necessity to go a certain way; for this "way-less" experience would be a "pseudo-familiarization" of the world (Anders, 1956, p. 117; cf. Schraube, 2009).

As if it was an acting subject, digital technology also seems to have emancipated itself from being merely a tool, an instrument, or a means to an end other than itself. It intrudes into the fabric of everyday life, kraken-like, looks for more and more applications to practice, and thus makes itself an indispensable part of human activities and societal ways of existence. It is not exclusion from, but rather inclusion in this process which appears as a problem. The apparatus reaches a new extension.

Consequently, this process obviously has many parallels with the problem of the "presumed inherent separability of observer and observed, knower and known" (see above). However, in what ways is the diffraction of the categories of subject–object useful in this current development? Returning to Lev S. Vygotsky, it can be shown that the epistemological problem cannot be resolved without dialectical theorizing, and becomes quite problematic with overstressing structural immanence, as can be found with Barad, and partly with Haraway.

## The Cyborg Metaphor Revisited—Or: The Manner of Doing Science

Looking back at Haraway's influential book of 1997, *Modest witness@ second millennium. FemaleMan meets OncoMouse: Feminism and Technoscience* (reprinted: Haraway & Goodeve, 2018), feminist philosophy celebrated a societal and paradigmatic shift which was perceived in the mirror of new technologies. Besides an ironic play of confusion, Haraway's essay revolved around questions of matter and science as it emerges in its 'situated' making. It was clarified that 'situated knowledge' should no longer be ignored and depreciated, so that scientific research gets a more sophisticated understanding of human action (cf. Suchman, 1987).

The postmodern critique as it was laid out by Haraway dared to break with the researcher's subjectivity invoked to 'witness' objectivity. This was seen as an abstraction disguising the concrete societal relations and practices of excluding subjectivities that were considered to be unsuitable for science (Haraway, 1985, p. 32). Therefore, Haraway's plea was to overcome women's exclusion from the production of new technologies and, similarly, to include all other persons concerned in the material development of their conditions of life.

However, although Haraway's critique was presented as a radical one, it is striking that the shortfalls of 'situated' knowledge as reflected (and not abstracted) concrete subjective experience did not attract the attention of postmodern and deconstructivist thinkers in similar ways. Barad's turn to structuralist immanence can be interpreted as a continuation within this move. As highlighted in the previous section however, digitalized environments demand scientific engagement with people's particular situatedness.

Looking at Barad's research program, it becomes obvious that there is a methodological reflection that is missing—with far-reaching consequences. Barad stresses the "intra-activity of the world" in an all-encompassing way, i.e., as a fundamental form by which "*matter comes to matter* through the iterative intra-activity of the world" (Barad, 2007, p. 152). But from the standpoint of an "iterative intra-activity of the world", the discourse in which we have learnt to articulate subjective reasons and responsibilities in relation to our actions and in relation to others is transformed into a discourse of rather impersonal processes. This disarticulates our particular engagements with societal practices *we are part of* and through which *we* produce and consume, for example, radioactivity, digitalization, etc. In a nutshell, methodologically, I suggest that this historical work of generations on our ways of knowing and experiencing ourselves cannot simply be revealed as an illusion or as an error, and thus, can be dismissed. The relation between the responsible 'agential' parts and their entire societal practice is not immediately available and criticizable against the backdrop of 'situated' ways of being and consciousness. However, if matter constantly intra-acts, often irrespective of human will, the individual faces uncertainty above all, and can merely hope that the ethically 'right' solution, or at least, the 'correct' socio-critical intervention in relation to the world will be ready to hand. Without developing human capacities to bring matter (both as societal and natural conditions) under control, we would endorse a fatalism with regard to the apparatus we are in.

## Scientification as a Politicized Capacity to Act

There is no epistemic rupture today with the thesis that the activities of thinking, knowing and recognition, including in a scientific manner, are time and field dependent (cf. Langemeyer, 2017b, p. 19). However, as the previous section shows, the mere acknowledgment of 'situated knowledge' and its rather global revaluation through a symmetry of matters is not a solution to the problem of scientists' and, more generally, humankind's responsibilities.

Against these shortcomings, the concept of "scientification" shall be strengthened. With Vygotsky, this approach agrees that the cognition of an individual cannot be scientific only in relation to itself. To establish a relation to certain *scientific concepts, methods* and *research results* in relation to capacities to act means participating in a certain domain of the historical practice of scientific thinking and knowledge production.

This is also true for the economic sectors where the production of technologies builds on scientific knowledge. Particularly, digitalization brings about a new level of scientification because it enables a close connection between the technological regulation of numerous processes and their mathematical operationalization. This means that, without science, the new digital 'universe' of information and automatic information processing would be disjointed, incoherent and as such useless. Digital data would be unusable for automated control if it did not 'incorporate' science.

However, scientification and technologization also increase the distance between the world of objects and the working subjects, so that their relationship becomes more indirect, more theoretical and thus opaque. The problems which IT workers, for instance, deal with are opaque and complex (e.g., software systems), and become intelligible only with activities to analyze, interpret, reveal, expand, experiment, test, and reinterpret the object of work (Langemeyer, 2017a).

Within research activities like these, workers may however be thrown back on believing and relying blindly on opaque technologies, on their 'interpretations', 'testing', 'calculations', and 'solutions' produced elsewhere. The capitalist relations of production create particular problems of distorted cooperation and transparency. A common phenomenon is therefore that particular work activities have become science-like activities: In this mode, subjects who are (or should be) *striving* for comprehension and agency in relation to unresolved problems are dependent on societal institutions that provide expertise (and sometimes pseudo-expertise) in testing, elaborating, and reconfiguring the matter which is to be brought under control (cf. Langemeyer, 2015, 2019a, b). These science-like activities are conducted individually and sometimes collectively. In either case, their potential mainly unfolds by overcoming the limits of distorted comprehension, which implies igniting the interaction between special and general science.

Science-like activities can therefore be interpreted as a transitory form of practice of a collective. This collective becomes more powerful through its engagements in overcoming the deficits of science-like activities. It strives for scientification by critically testing and theorizing in depth why capacities to think are incomplete or imperfect to see the broader picture of interrelations (ibid.). This transition from science-like to scientificated practices includes reconfiguration, communication, debate, and further investigation to ensure correct, precise and appropriate thinking and reasoning.

Politically, economically and scientifically, the precarious scientification process is therefore essential and needs societal conditions for independence. Its precariousness is thus entangled with many societal relations of time and space, which might be simultaneously economic, political, cultural, etc. And in these challenging entanglements, critical perspectives are to be generated—each time anew. The need for critique emerges as situated in practice, but the presuppositions and the capacities for thinking and acting in a critical manner depend on long-term cultural development.

## Concluding Remark

The question can be raised whether this approach is, intentionally or unintentionally, inclined to rationalist and scientistic positions since it accepts continuing with subject and object as poles in the epistemic process, and with referring responsible and conscious activity to humans only. This criticism would be misleading. I assume that there is no guarantee of exceeding the numerous science-like activities through scientification. Within science-like activities, the different positions of researcher subjects are not unproblematic. They are concrete societal conditions under which the objects of study are identified, interpreted, and construed. I agreed with Haraway and Barad that moving to more comprehensive forms of knowledge must be an engagement with the world, a way of taking on responsibility for it, but to become more scientific this must be combined with self-critical engagements. The sources for developing these engagements further are social, and from there, become individual. They are social and thus material, yet we should not resign ourselves to structural immanence. Raising consciousness is subjective activity and not intra-activity. The scientification approach therefore emphasizes that, also in scientific practice, individuals must take a stance and then take on the responsibility for this in relation to concrete others and future generations. This is what distinguishes their scientific engagement (as sociohistorical human practice) from matter like radioactivity or digital data. Ultimately, it is this responsibility which seems to be diverted by "agential realism"—an awkward peripeteia, especially with regard to current technological developments.

## References

Anders, G. (1956). *Die Antiquiertheit des Menschen. Teil 1*. München: C.H. Beck.

Althusser, L. (1962). Le«Piccolo», Bertolazzi et Brecht: Notes sur un théâtre matérialiste. *Esprit, 312*(12), 946–965.

Althusser, L. (2006). *Philosophy of the encounter: Later writings, 1978–87*. London: Verso.

Althusser, L. (2014). *On the reproduction of capitalism: Ideology and ideological state apparatuses*. London: Verso (Orig. French 1970).

Bachelard, G. (2006). *The formation of the scientific mind: A contribution to a psychoanalysis of objective knowledge*. Manchester: Clinamen Press (Orig. French 1934).

Barad, K. (2003). Posthumanist performativity: Toward an understanding of how matter comes to matter. *Signs: Journal of Women in Culture and Society, 28*(3), 801–831.

Barad, K. (2007). *Meeting the universe halfway: Quantum physics and the entanglement of matter and meaning*. Durham and London: Duke University Press.

Barad, K. (2011). Erasers and erasures: Pinch's unfortunate 'uncertainty principle'. *Social Studies of Science, 41*(3), 443–454.

Barad, K. (2017). Troubling time/s and ecologies of nothingness: Re-turning, re-membering, and facing the incalculable. *New Formations a Journal of Culture/Theory/Politics, 92*(1), 56–86.

Brandstätter, V., & Otto, J. H. (2009). *Handbuch der Allgemeinen Psychologie: Motivation und Emotion*. Göttingen: Hogrefe.

Canguilhem, G. (2002). *Écrits sur la médecine*. Paris: Seuil.

Friedrich, J. (2012). *Lev Vygotski: médiation, apprentissage et développement: une lecture philosophique et épistémologique*. Geneva: Carnets des sciences de l'education.

Haraway, D. (1985). A manifesto for cyborgs: Science, technology, and socialist feminism in the 1980s. *Socialist Review, 80,* 65–107.

Haraway, D. (1992). The promises of monsters: A regenerative politics for inappropriate/d others. In L. Grossberg, C. Nelson, & P. Treichler (Eds.), *Cultural studies* (pp. 295–337). New York: Routledge.

Haraway, D. J., & Goodeve, T. (2018). *Modest_Witness@ Second_Millennium. FemaleMan_Meets_OncoMouse: Feminism and technoscience*. London: Routledge (Orig. 1997).

Heider, F. (1958). *The psychology of interpersonal relations*. New York: Wiley.

Jaeggi, R. (2005). *Entfremdung. Zur Aktualität eines sozialphilosophischen Problems.* Frankfurt a. M.: Campus.

Knorr Cetina, K. (2007). Knowledge in a knowledge society: Five transitions. *Knowledge, Work & Society, 4*(3), 25–42.

Kucklick, C. (2014). *Die granulare Gesellschaft: Wie das Digitale unsere Wirklichkeit auflöst*. Berlin: Ullstein eBooks.

Kuhl, J., & Kaschel, R. (2004). Entfremdung als Krankheitsursache: Selbstregulation von Affekten und integrative Kompetenz. *Psychologische Rundschau, 55*(2), 61–71.

Langemeyer, I. (2015). *Das Wissen der Achtsamkeit: Kooperative Kompetenz in komplexen Arbeitsprozessen* [Engl.: The knowing of mindfulness: Cooperative competence in complex work processes]. Münster: Waxmann .

Langemeyer, I. (2017a). Methodological challenges of investigating intellectual cooperation, relational expertise and transformative agency. *Nordic Journal of Vocation Education and Training, 7*(2), 39–62.

Langemeyer, I. (2017b). The field concept in psychology, Gestalt theory, physics, and epic theatre—Brecht's adaptations of Kurt Lewin. *Journal of New Frontiers in Spatial Concepts, 9*(1), 1–16.

Langemeyer, I. (2019a, in press). Mindfulness in cooperation and the psychodynamics in high-reliability-organizations. *Annual Review of Critical Psychology, 15*.

Langemeyer, I. (2019b). Philosophical foundations of adult education. In M. A. Peters (Ed.), *Encyclopedia of educational philosophy and theory.* New York: Springer. https://doi.org/10.1007/978-981-287-532-7_671-1.

Langemeyer, I., & Roth, W.-M. (2006). Is cultural-historical activity theory threatened to fall short of its own principles and possibilities as a dialectical social science? *Outlines: Critical Social Studies, 8*(2), 20–42.

Marx, K. (1947). Theses on Feuerbach. In K. Marx & F. Engels (Eds.), *The German ideology* (pp. 121–123). New York: International Publishers (Orig. German 1845).

Ratner, C., & Silva, D. N. H. (Eds.). (2017). *Vygotsky and Marx: Toward a Marxist psychology*. London and New York: Taylor & Francis.

Schraube, E. (2009). Technology as materialized action and its ambivalences. *Theory and Psychology, 19*(2), 296–312.

Sève, L. (2018). Où est Marx dans l'œuvre et la penseée de Vygotski? In *7e Séminaire International Vygotski, Génève* (manuscript).

Stetsenko, A. (2016). *The transformative mind: Expanding Vygotsky's approach to development and education*. Cambridge: Cambridge University Press.

Suchman, L. A. (1987). *Plans and situated actions: The problem of human-machine communication*. New York, NY: Cambridge University Press.

Urhahne, D. (2008). Die sieben Arten der Lernmotivation. Ein Überblick über zentrale Forschungskonzepte. *Psychologische Rundschau, 59,* 150–166.

Vygotsky, L. S. (1978). *Mind in society: The development of higher psychological processes*. Cambridge, MA: Harvard University Press.

Vygotsky, L. S. (1987). Thinking and speech. In R. W. Rieber & A. S. Carton (Eds.), *The collected works of Vygotsky vol. 1: Problems of general psychology* (pp. 39–285). New York: Plenum (Orig. Russian 1934).

Vygotsky, L. S. (1997). The historical meaning of the crisis in psychology: A methodological investigation. In R. W. Rieber & J. Wollock (Eds.), *The collected works of LS Vygotsky* (pp. 233–343). New York: Springer US (Orig. Russian 1927).

# Index





K. C. O'Doherty et al. (eds.), *Psychological Studies of Science and Technology*, Palgrave Studies in the Theory and History of Psychology,
https://doi.org/10.1007/978-3-030-25308-0

I



K



L



M



N